Stress and the Developing Brain

Colloquium
Digital Library of Life Sciences

This e-book is a copyrighted work in the Colloquium Digital Library—an innovative collection of time saving references and tools for researchers and students who want to quickly get up to speed in a new area or fundamental biomedical/life sciences topic. Each PDF e-book in the collection is an in-depth overview of a fast-moving or fundamental area of research, authored by a prominent contributor to the field. We call these e-books *Lectures* because they are intended for a broad, diverse audience of life scientists, in the spirit of a plenary lecture delivered by a keynote speaker or visiting professor. Individual e-books are published as contributions to a particular thematic **series**, each covering a different subject area and managed by its own prestigious editor, who oversees topic and author selection as well as scientific review. Readers are invited to see highlights of fields other than their own, keep up with advances in various disciplines, and refresh their understanding of core concepts in cell & molecular biology.

For the full list of published and forthcoming Lectures, please visit the Colloquium homepage:
www.morganclaypool.com/page/lifesci

Access to Colloquium Digital Library is available by institutional license. Please e-mail info@morganclaypool.com for more information.

Morgan & Claypool Life Sciences is a signatory to the STM Permission Guidelines. All figures used with permission.

Colloquium Series on The Developing Brain

Editor
Margaret M. McCarthy, PhD,
Professor and Chair
Department of Pharmacology
University of Maryland School of Medicine

The goal of this series is to provide a comprehensive state-of-the-art overview of how the brain develops and those processes that affect it. Topics range from the fundamentals of axonal guidance and synaptogenesis prenatally to the influence of hormones, sex, stress, maternal care, and injury during the early postnatal period to an additional critical period at puberty. Easily accessible expert reviews combine analyses of detailed cellular mechanisms with interpretations of significance and broader impact of the topic area on the field of neuroscience and the understanding of brain and behavior.

My research program focuses on the influence of steroid hormones on the developing brain. During perinatal life, there is a sensitive period for hormone exposure during which permanent cytoarchitectural changes are established. Males and females are exposed to different hormonal milieus and this results in sex differences in the brain. These differences include alterations in the volumes of particular brain nuclei and patterns of synaptic connectivity. The mechanisms by which sexually dimorphic structures are formed in the brain remains poorly understood.

I received my PhD in Behavioral and Neural Sciences from the Institute of Animal Behavior at Rutgers University in Newark, NJ in 1989. I then spent three years as a post-doctoral fellow at the Rockefeller University in New York, NY and one year as a National Research Council Fellow at the National Institutes of Health, before joining the faculty at the University of Maryland. I am a member of the University of Maryland Graduate School and the Center for Studies in Reproduction. I am also a member of the Society for Behavioral Neuroendocrinology, the Society for Neuroscience, the American Physiological Association, and the Endocrine Society.

Stress and the Developing Brain
Lisa Wright and Tara Perrot
www.morganclaypool.com

ISBN: 9781615045273 paperback

ISBN: 9781615045280 ebook

DOI: 10.4199/C00069ED1V01Y201211DBR009

A Publication in the

COLLOQUIUM SERIES ON THE DEVELOPING BRAIN

Lecture #9

Series Editor: Margaret M. McCarthy, University of Maryland School of Medicine

Series ISSN

ISSN 2159-5194 print

ISSN 2159-5208 online

Stress and the Developing Brain

Lisa Wright and Tara Perrot
Department of Psychology & Neuroscience
Dalhousie University

COLLOQUIUM SERIES ON THE DEVELOPING BRAIN #9

MORGAN & CLAYPOOL LIFE SCIENCES

ABSTRACT

The human brain does not develop in a vacuum according to a set of predetermined blueprints—it is involved in a dynamic interplay with the environment that influences gene expression and ultimately structure and function. Some cortical regions, such as the prefrontal cortex (PFC) undergo structural changes throughout the adolescent period and into early adulthood, making their structure and functions particularly interesting to study with respect to gene-environment interactions. Repeated exposure to stress is a predisposing factor in the emergence of various mental illnesses, such as anxiety and depression, although this is by no means an absolute relationship. While some people appear to be vulnerable to the effects of repeated stressors, others are resilient, and this individual variability is partly due to developmental programming of brain regions involved in modulating stress responding, such as the PFC. In the present book, we will discuss features of adolescent brain development that may provide a basis for neural plasticity in stress responding: the highly protracted development of the PFC, the profound change in interconnectedness among cortical and subcortical brain regions, and the characteristic 'rise and fall' pattern for many of the late-developing aspects of neural architecture in PFC and other stress-related brain regions.

KEYWORDS

stress, adolescent, human, prefrontal cortex, development, developmental programming, animal models, early experience, epigenetics

Contents

CHAPTER 1

The Stress Response System

1.1 INTRODUCTION

The age-old debate of genes versus environment has finally been put to rest as it has become abundantly clear that adult phenotype is the product of an intricate interplay between genetic code and environmental modulation of gene expression throughout the lifespan, but in particular during periods of developmental plasticity (Fig. 1). The move from considering 'genes vs. environment' to instead considering 'environmental modulation of gene expression' has been aided by the growth of the field of epigenetics (literally 'around genetics'; indicating gene expression alterations that result in changes to physiology and behaviour, but do not involve alterations to gene sequence) within the fields of neuroscience and neuroendocrinology. The idea of an epigenetic landscape existing in which gene regulation modulates development was first put forth by developmental biologists studying anatomical structure [1] and has more recently been applied at the cellular level to the developmental fate of stem cells [2]. The general concept is useful in discussions of nervous system development during which we know that events can exert permanent effects on adult function and expression of behaviour. In general, the idea is that developmental information does not pre-exist within the immature organism but is constructed as development progresses. This is reflected in a change in thinking from 'pre-determined epigenesis' to 'probabilistic epigenesis,' as outlined in Figure 2A. The dynamic interaction between genes, brain structure/function/behaviour, and experience provides the opportunity for several phenotypes to emerge depending on the paths taken during development. The phenotype of an organism is the set of expressed physical and behavioural characteristics that result from the interaction of environmental influences on gene expression. As development progresses and paths are embarked upon, there are fewer opportunities for changing them, consistent with the idea that plasticity is reduced as development progresses. This is schematically represented in Figure 2B as a modified Plinko game. In the game of Plinko, balls or discs are dropped randomly from the top and the eventual path they take depends on their interactions with the barriers in place along the way. Similarly, as development progresses, genes are turned off or on in the organism, shaping the phenotype. Some of these gene inactivation or activation events occur at particular points in development. Thus, organisms with similar genetic sequences etch out different phenotypic paths, depending on the environmental conditions encountered along the way,

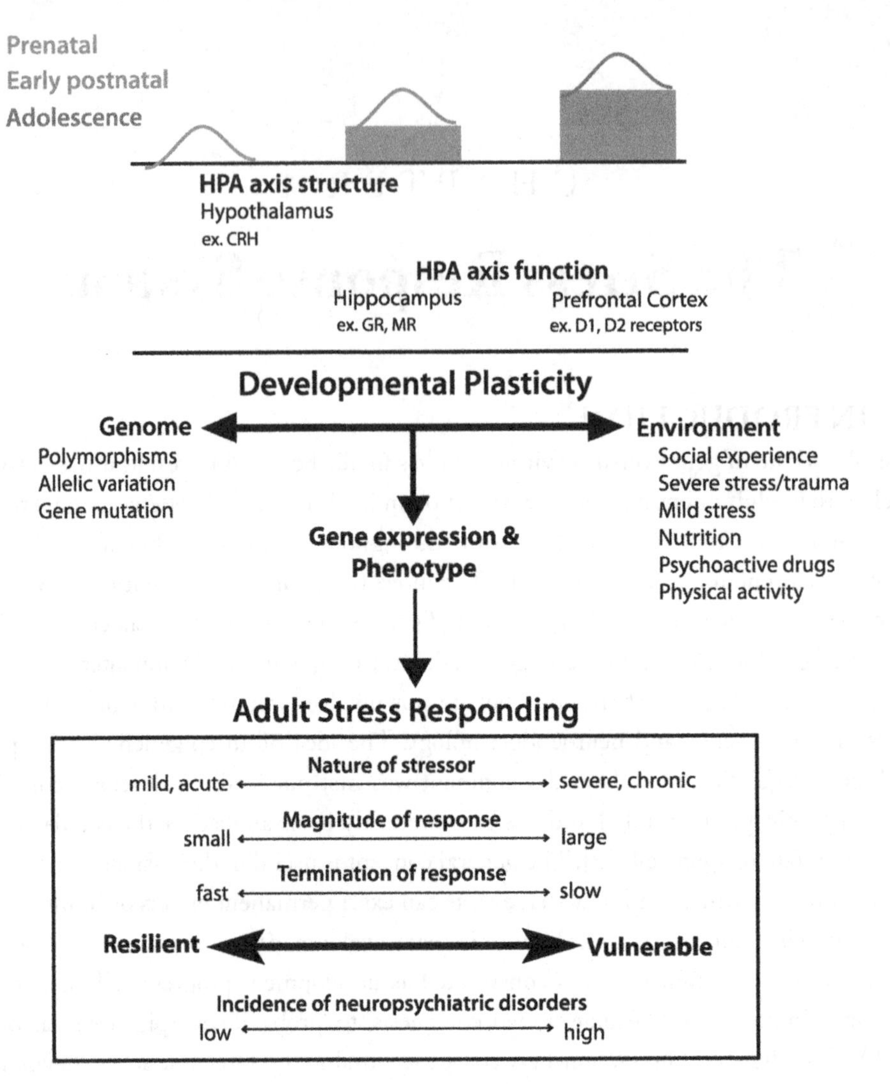

FIGURE 1: Adult stress responding is variable, depending on factors such as the nature of the stressor, among others. The magnitude of the response to a stressor, and the efficiency with which the response (which is energetically costly) is terminated, vary among individuals, placing them at different locations on a continuum between resilient and vulnerable. Vulnerable individuals are those for whom stress responding is more tightly associated with the development of neuropsychiatric disorders, such as anxiety and depression. Where an individual ends up being on this continuum is a result of gene expression that is dictated by interactions between components of the individual's genome and aspects of the environment. This developmental plasticity is most salient during particular sensitive periods (i.e. prenatal, early postnatal, adolescent), each of which represents a time when different components of the major stress system, the hypothalamic pituitary adrenal (HPA) axis, are being formed. During these sensitive periods (illustrated by a coloured wave), the underlying brain regions, hormones and neurotransmitter systems associated with stress responding are programmed. As indicated by the coloured blocks, different aspects are programmed during these different developmental periods, and events programmed during ensuing periods occur on the foundation of those components previously programmed.

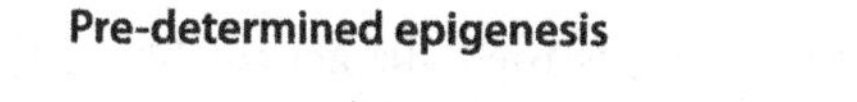

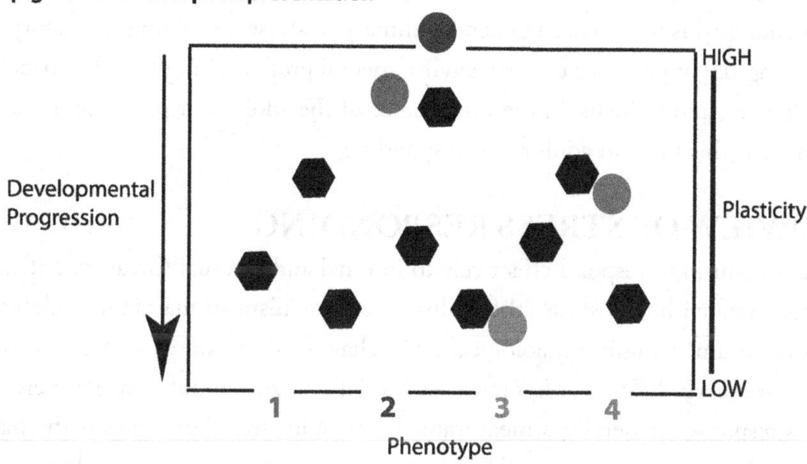

FIGURE 2: (A) Traditionally, the view of how individuals attain the experience they do was of a predetermined set of genes being expressed as a result of the genetic code, which would in turn dictate brain structure and function, and, by virtue, our behaviour and experience. With what has been gleaned by developmental biologists over the last few decades, a new view has emerged that takes into account the dynamic way in which gene expression is modified by our experience. We are now much more aware that what is contained within our genome is simply a probabilistic blueprint for the potential expression of genes, and thus, our phenotype, the sum product of all of the physical and behavioural attributes we come to express as adults. (B) As development progresses and particular paths (developmental trajectories) are embarked upon, there are fewer opportunities for changing them, consistent with the idea that plasticity is reduced as development progresses. This is schematically represented here as a slightly modified plinko game. In the game of plinko, balls or discs are dropped randomly from the top and the eventual path they take depends on their interactions with the barriers in place along the way. In this game, as the balls get further down the board, there are more barriers (hexagons) and thus, the path becomes more constrained. Similarly, organisms with similar genetic sequences etch out different phenotypic paths depending on the environmental conditions encountered along the way. As development progresses, genes are turned off or on in the organism, shaping the phenotype (represented by different coloured numbers). Just as the path becomes more constrained later in the plinko game by more barriers, phenotypic plasticity (flexibility or variation) decreases over time.

particularly during critical periods. In our version of the game, as the balls get further down the board, there are more barriers representing developmental events, and thus the path becomes more constrained and phenotypic plasticity (flexibility or variation) is reduced.

Developmental programming of adult stress responding is an excellent model system in which to study the inseparability of genes and environment. Indeed, the study of prenatal and early postnatal programming of adult stress responding has been in full swing for many years, with particular emphasis on regulation by the parts of the brain that are complete in their development early in life. What has been less studied has been the role that later developmental periods, and later developing brain regions, play in influencing the way in which the adult animal responds to stressors. In this book, our goal is to provide a general summary of stress responding, a cursory review of the research showing the importance of early environmental programming (this has been reviewed many times elsewhere), and to focus on the importance of the adolescent period and brain regions under development at this time on adult stress responding.

1.2 OVERVIEW OF STRESS RESPONDING

The ability of an organism to respond effectively to internal and external threats is critical for survival. Events that threaten homeostasis (the ability of an organism to maintain a relatively stable physiological equilibrium) require physiological and behavioural responses in order to regain homeostasis and ensure survival. Stressor is a term used to refer to any stimulus, real or perceived, that results in a stress response. Generally, a measurable increase in stress hormones is the hallmark of a stressor response; however, there is good evidence that such hormones are involved more in the recovery from a stressful event and not in response to the stressor *per se* [3]. A more comprehensive discussion of defining a stress response is beyond the scope of this chapter (but see [3]). Although we will describe the hormones released in response to a stressor, we will focus on behavioral responding to stimuli perceived to be stressful and argue that a measurable change in behavior is most important. For example, responses to even mild stressors, such as exposure to heat, include a behavioral response (ex. panting or shade-seeking).

It is worth noting here that there are multiple systems in place that help to maintain homeostasis in response to threat. These are activated to differing extents depending on the timing, the severity, and the nature of the stressor. Below, we will summarize the basics of two of these systems that have particular relevance to human stress responding.

1.2.1 Sympathoadrenal System (SAS)

The autonomic nervous system is involved in regulating a variety of bodily functions through connections of nerves to and from the spinal cord to organs such as the heart and intestines. It is divided into the sympathetic and parasympathetic branches, which mediate vigorous and vegetative functions, respectively. Activation of the sympathetic branch prepares organ systems for a 'flight or fight' response to a threat, increasing heart rate and decreasing digestive function, for example. The sympathetic branch also has connections to the medullary portion of the adrenal gland and this comprises the sympathoadrenal system (SAS). Activation of the sympathetic system results in the release of hormones, norepinephrine (NE) and epinephrine (Epi), from the adrenal medulla (Fig. 3). The physiological effects of these hormones are vast but in general, they subserve short-term adaptive functions. For example, NE plays a prominent role in stress-induced activation of a brain region called the amygdala, inducing a hyper-vigilant state that enables the organism to recall the event [4]. Interestingly, this memory is further enhanced if the stressor is sufficient to stimulate the amygdala with slower-acting hormones released by the adrenal cortex (discussed below [4]), demonstrating the interplay between various stress systems in mediating adaptive responses. Furthermore, in terms of neural stress responding, entire new systems of regulation over gene expression are currently being discovered, such as the involvement of noncoding microRNAs [5].

1.2.2 Hypothalamic-Pituitary-Adrenal (HPA) Axis

A full review of the HPA axis and other components of the vertebrate stress response system is not possible here, and the reader is referred to Tsigos & Chrousos, 2002 [6]. Here, we summarize the major points. Activation of the HPA axis upon perception of a stressor is a key component of the vertebrate stress response system (Fig. 3). Stress-responsive neurons in the brain detect the threatening stimulus and initiate a cascade of hormones, which begins with the release of corticotrophic releasing hormone (CRH) from the paraventricular nucleus (PVN) of the hypothalamus and results in the eventual release of several hormones from the adrenal gland into the circulation. These hormones travel in the bloodstream to a number of target organs and exert effects on various physiological systems, such as those regulating heart rate, blood glucose and protein levels, and perhaps most importantly, effective behavioural responses. Some of the immediate effects are not mediated by the HPA axis but by activation of the sympathetic nervous system, as discussed above. This results in the release of hormones from the adrenal medulla, and these exert short-term adaptive effects that are classically referred to as being part of the 'flight or fight' response to a threat. Other effects are on a longer-term basis, allowing the organism to not only deal with the immediate threat but to also recover from the effects of the stressor on the body. Many of these recovery effects are mediated by release of glucocorticoids (GCs) from the adrenal cortex. Even the responses mediated by GC release alone are heterogeneous and occur at various levels (molecular, cellular, behavioral) and across time from rapid to slower changes. Such a variety of responses are possible because of the intricate interplay of effects mediated by the receptors activated by GCs and other key players, such as steroid binding proteins (see [7]). Following release, GCs act on two receptor subtypes: glucocorticoid receptors (GR) and mineralocorticoid receptors (MR), which differ in their affinity

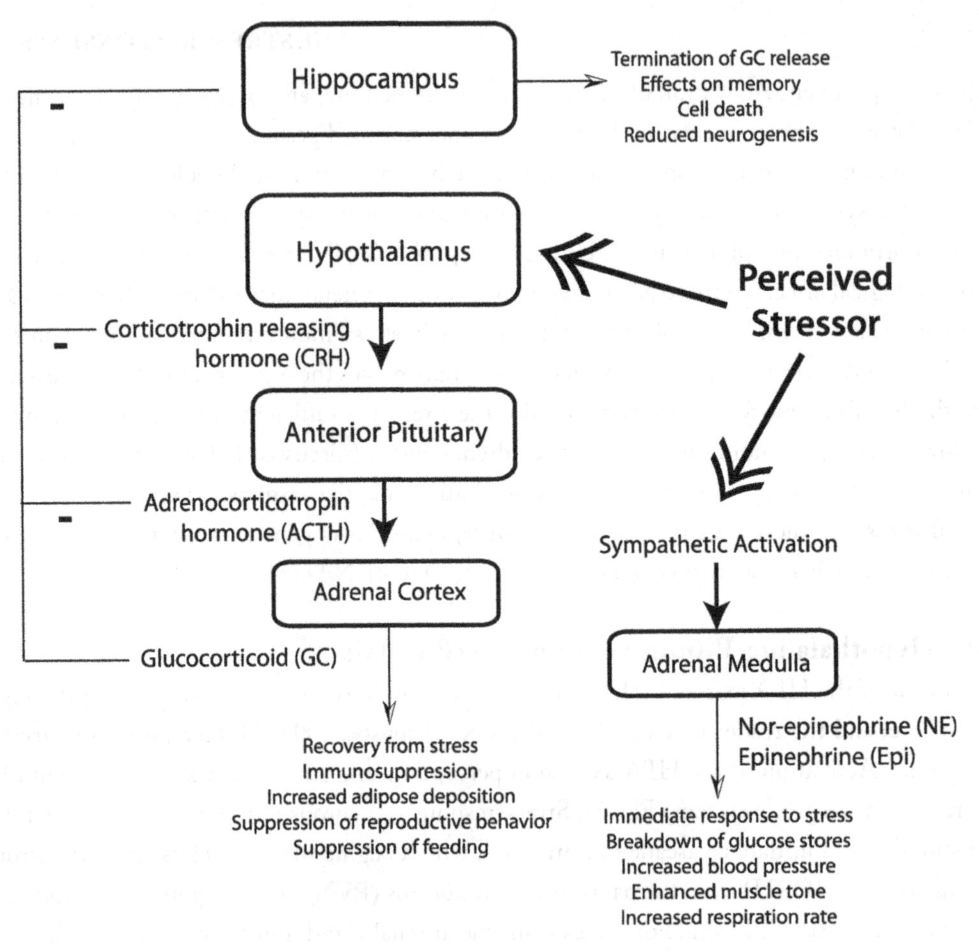

FIGURE 3: Following perception of a stressor, a number of responses are initiated that differ in terms of how quickly they are activated and how long they last. Very quickly, sympathetic nervous system activation of the adrenal medulla results in the release of norepinephrine (NE) and epinephrine (Epi) into the bloodstream, which exert actions on various organ systems to prepare the body for dealing with an immediate threat. Perception of a stressor also activates the hypothalamus and results in the release of corticotrophin releasing hormone (CRH) from the paraventricular nucleus. In addition to exerting its own effects on behavior by acting on receptors in various parts of the brain, CRH acts on receptors in the anterior pituitary gland to cause the release of adrenocorticotrophin hormone (ACTH). By the time that ACTH acts on the adrenal cortex to release glucocorticoids (GCs), minutes have passed and in many cases, GCs aid in the recovery phase following exposure to acute stressors. Longer-term stressors are deleterious to the organs on which GCs act (most of them, including a number of parts of the brain) because they are catabolic. They suppress a number of generally adaptive behaviours, such as feeding, and in general, repeated and/or chronic stressor exposure is bad for one's health. GC exposure is terminated by negative-feedback loops that are initiated when GCs bind to receptors at all levels of the HPA axis; however, most of the termination is afforded by actions initiated through GC binding to receptors in the hippocampus. A number of negative effects to hippocampal structure/function are observed with excess GC exposure.

for binding GCs. At low circulating concentrations of GCs, binding of MR occurs to a point of saturation, after which GR become occupied. In general, MR appear to mediate the more rapid effects of GCs involved in increasing cellular excitability and underlying increased alertness, arousal and vigilance at the behavioral level [8]. As GR become bound with higher levels of GCs, the system becomes tuned to recovery, including normalization of cellular excitability and consolidation and preparation for future behavioral responding [7, 8].

Stressors that signal immediate threats to survival, such as heat or hunger, can activate the HPA axis in a 'reactive' manner, a process mediated by direct projections of ascending brain systems or circumventricular organs onto the PVN [9]. However, most of the stressors encountered by humans in the developed world are of an 'anticipatory' nature—although there is no immediate threat to homeostasis, the HPA axis is activated to prepare the organism for the potential of danger. These types of stressors are commonly referred to as 'psychogenic stressors' and their ability to activate the HPA axis involves brain regions that have the capability to integrate information about the external environment (ex. smell, sound) with innate programs (ex. smell represents a predator) or with memories (ex. the last time I was foraging here, there was a predator). Such brain regions include limbic and association areas and indirect connections with the PVN [9]. Limbic areas most relevant to the present review include the hippocampus and prefrontal cortex (PFC), both of which are involved in anticipatory responses to stressors and are implicated in dysregulation of the HPA axis in response to chronic activation (see Section 1.2.2).

The hippocampus is a limbic brain region that contains high concentrations of both MR and GR, providing it with the opportunity to respond to a wide range of GC concentrations, including those that would be circulating following stressor exposure. Numerous pieces of evidence suggest that the hippocampus plays a key role in negatively regulating the HPA axis [9]. In particular the ventral portion of the hippocampus appears critically important as indicated by the fact that lesions of the ventral hippocampus increase CRH levels in the PVN and enhance HPA axis activation in response to a stressor; however, this may not be the case with respect to responding to chronic stress [10, 11].

1.2.3 Stress-Related Behaviours

One of the challenges in the field of stress research is defining the stress response. What response must be present for a researcher to consider the subject 'stressed'? Must there be measurable levels of GCs in the blood? What if there is a behavioral response, but GCs aren't elevated? Is that sufficient for the stimulus to be considered 'a stressor' and the response to be considered a 'stress response'? This debate is ongoing in the field [12] and a nice summary of it can be found elsewhere, in which the authors conclude that both levels of GCs and behavioral responses are 'stress metrics' [3]—quantifiable changes that can be used to estimate activation of the HPA axis. We will argue throughout this book that behavioral responses to stressors, no matter how simple, are extremely important and measuring behavioral responses should always be attempted.

Ultimately, the behaviors displayed by an organism in response to threat will be the most important factor determining survival. The behavioural response initiated by a stressor will depend on the species, the type and duration of stressor exposure, environmental context, as well as variables that are specific to the organism such as sex and reproductive status, and developmental stage. Some generalizations can be made, however, especially with respect to basic behaviors such as reproduction, foraging, and defensive behavioral actions. Research has been performed in the wild and in the laboratory and both have their advantages/disadvantages. Research on the effects of chronic stress in the wild have the advantages of ecological validity with respect to the stressor and the responses. They lack the controlled setting of the lab, however, and suffer from the disadvantages associated with potential methodological confounds. Nonetheless, studies from the wild have been invaluable in demonstrating that basic innate behaviors can be severely affected by ecologically relevant stressors, such as predation threat. Under predation risk, various species of wild rodents have been shown to display changes in feeding behavior, reproduction-related behaviors, decreased pain sensitivity, changes in activity levels, and over the longer term, significant effects on evolutionary fitness, by negatively impacting mating behavior and timing of breeding [13, 14, 15].

In the laboratory situation, stress-related behavioral responses are limited in a number of ways. First, the stressors commonly used in the laboratory, such as restraint/immobilization, foot shock, or swim stress, either do not allow for measurable behavioral responses (ex. restraint), or the responses that they do evoke are subject to a heavy dose of interpretation. For example, forcing rodents to swim for 10 min in deep water with no means of escape is sufficient to elevate GC levels [16]. This stressor also induces various behaviors, such as struggling to escape, but also eventual immobility. The amount of immobility (or latency to engage in the behavior) is often used as a marker of 'behavioral despair' [17]; the idea being that an animal that becomes immobile sooner and for longer has effectively given up. The problem is that one could argue that becoming immobile in an inescapable situation is adaptive by saving energy, and animals that adopt that behavioural strategy sooner are simply showing more adaptive behavior. This difference in potential interpretation of behavior underscores the need for studying behavioral responses to stressors that an organism might encounter and have to deal with in its natural environment. The use of ecologically relevant stressors increases the chances that responses will be ecologically relevant as well. We routinely use the odor from a predator (ex. cat) as an ecologically relevant stressor for lab rats. Using this stressor and a simple open field, we can measure a number of behaviors including avoidance of the odor and reductions in activity levels, behaviors that are easily quantifiable and consistently observed. By adding a place to hide within the open field, we can also examine risk assessment behaviors, such

VIDEO 1: http://tinyurl.com/wrightperrot-video1 Footage of a laboratory rat in the open-field testing apparatus we use for measuring defensive behaviour. The rat is behind a hide box and a piece of cat collar impregnated with cat odour is affixed to the opposite end of the apparatus. As can be seen in the video and described in the text in Section 1.2.3, the rat displays a number of behaviours (ex. head-out from behind barrier) that are designed to assess the threat while still maintaining a protective vantage point. This footage was collected as part of a study that had been approved by the University Committee on Laboratory Animals at Dalhousie University using guidelines set out by the Canadian Council on Animal Care.

as the stretch-attend posture (rat stretches its body flat from within the hiding place) and head-out posture (rat sticks the snout or head out from within the hiding place). As can be seen in Video 1, such behaviors serve to maintain a certain level of safety while providing a protected scan of the threat. Figure 4 shows a schematic of the testing apparatus used in Video 1 and also summarizes data for head-out frequency across repeated exposures of male and female rats to the odor of a cat.

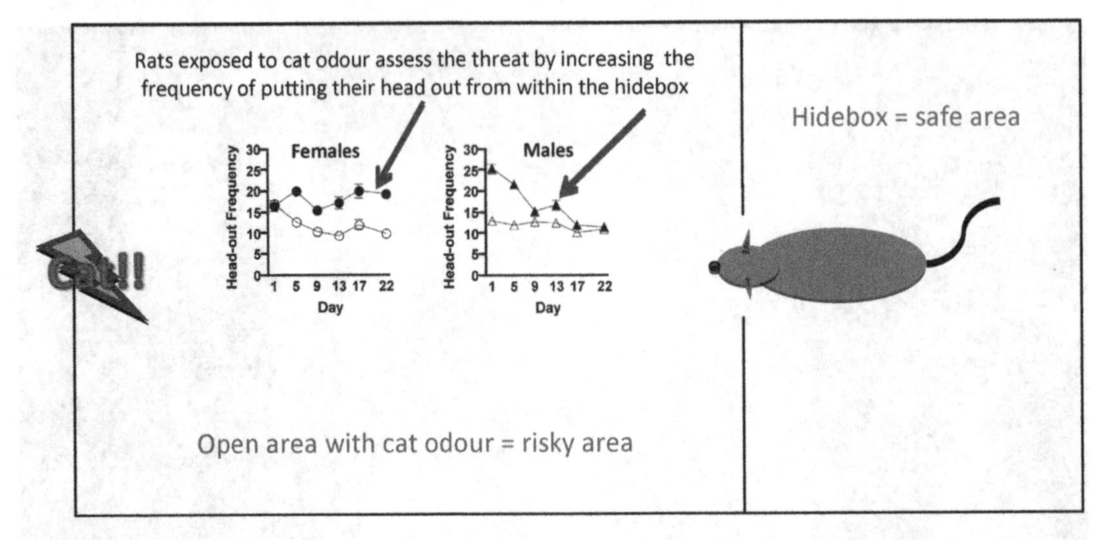

FIGURE 4: A schematic diagram of a stressor exposure arena used to assess behavioural measures of stress responding in rats. The exposure arena is equipped with a hidebox located at the end opposite from the stressor stimulus, which is a piece of worn cat collar that exudes the odors of a cat. To escape from the stressor, exposed rats will limit their time in the middle of the arena (the risky area) and will instead retreat into the hidebox, where specific defensive behaviours can be quantified, such as the time spent hiding and the frequency of surveying the environment by poking the head out of the hidebox door (i.e., head-out frequency). An example of head-out frequency data was taken from [22] is superimposed over the image of the arena. Closed symbols—mean (plus standard error of the mean) frequency for rats exposed to cat odour; Open symbols—mean (plus standard error of the mean) frequency for rats exposed to a control odour (piece of cat collar never worn by a cat).

The HPA axis evolved in conjunction with other systems to enable an organism to deal with short-term stressors that threaten homeostasis. Activation of systems by stimuli that threaten homeostasis results in the coordination of behaviors that are necessary for dealing with the threat, and this further disrupts homeostasis in the short-term, but once the threat is adequately dealt with, homeostasis is then restored by further coordination of behaviours. Such behaviors span a continuum of complexity from simple ones, such as shivering in response to cold, to foraging for food in response to hunger, to evasive maneuvers in response to a predation threat. The underlying neurobiological mechanisms controlling such behaviors and how they are initiated by the neuroendocrine cascades that occur in response to a stressor are not well understood. It has been possible to elucidate the brain regions activated by specific stressors, but causally connecting that activation to the endocrine events that occur in response to the threat has proven more difficult. For example, it has been known for a number of years that exposure of rats to the odor of a predatory species, such as a cat, results in activation of defensive behaviors, such as freezing, avoidance, and vigilance (e.g., stretch-attend postures). The neural circuitry underlying such behavioral responses has been mapped out using various techniques [18]; however, the means by which stress hormones interact with the neural circuitry to mediate the known behavioral output is far from clear. However, we have conducted a number of studies over the past several years to characterize stress responses of rats to predator odor across adolescent development and also to determine whether or not adolescent experience shapes adult behavioral and hormonal stress responses [19, 20, 21, 22]. Interestingly, it has been demonstrated that although there is significant release of stress hormones in response to predator odor, it is often dissociated from the behavioral responding [22, 23]. These findings suggest that some of the more newly discovered acute responses to stress, such as increased transcription of particular, non-protein-coding microRNAs [5], may play a more direct role in mediating immediate behavioral responses, whereas stress hormones may be more important for integrating information and coordinating activity in the various systems involved in generating behavioral output and reinstating homeostasis. All of theses mechanisms may be involved in epigenetically programming the adult phenotype during adolescent development, and further studies should be designed to further elucidate these processes.

1.2.4 Pathophysiological Consequences of Stress Responding

A number of conditions can precipitate a situation under which HPA activation ceases to mediate adaptive behavioral responses that allow an organism to effectively deal with stress. Chronic or repeated activation of the HPA axis, or a lack of control over acutely stressful situations, results in non-adaptive behavioral responding [7]. The question is whether non-adaptive stress responding is associated with disease processes—those involving organs outside the central nervous system (CNS), as well as those considered to be neuropsychiatric.

There is strong prospective evidence to suggest a link between physiological stress responding and coronary conditions, such as hypertension [24] and carotid atherosclerosis [25], both of which go on to predict future disease. Although dysregulation of the HPA axis has been suggested to be an etiological link between reported increases in stressful life events and development of various psychopathologies, such as depressive disorder [26], this research is based mostly on circumstantial evidence from cross-sectional studies. Indeed, much of this work relies on the fact that the link between stress and neuropsychiatric disorders is variable. Not every person exposed to even severe trauma develops a neuropsychiatric disorder. Thus, it is possible to examine the range of normal variation in stress responding, and link it to the development of neuropsychiatric disorders.

1.2.5 Sex Differences

Given the variability in stress responding and its consequences, a major goal in this research area is to uncover the factors that contribute to such variability. One such factor is an organisms' biological sex. While many prefer to use the term 'gender difference' when referring to men versus women, we will use the term 'sex difference' here because we will discuss literature from species in addition to humans, and the term 'gender' encompasses not only biological sex but also one's perception of their sexuality. The latter is not routinely assessed in studies in which differences between males and females in stress responding are measured.

Sex differences in stress responding are far from clear-cut, and some excellent reviews have been written summarizing the complexity (see [27, 28]). A general comment that can be made of the situation is that the direction of differences between males and females depends on the nature of the stressor and the stress measure quantified. However, consistent effects of sex, menstrual cycle phase, and menopausal status can be observed in response to acute psychosocial stress [28]. During reproductively active years, women show lower autonomic and HPA axis responses to such stress than men; however, during the luteal phase of the menstrual cycle (when female hormone levels are low), women show GC levels more similar to men in response to acute psychosocial stress. Pregnancy tends to dampen stress responding, while during menopause, women show increased SAS and HPA activity. Both are likely associated with altered gonadal hormone levels, as menopausal oral hormone replacement therapy attenuates the increase, and estrogen has known dampening effects on SAS responsiveness [28]. Nonetheless, the magnitude of the difference in the neuroendocrine responses to stress is relatively small, and sex differences in the consequences to stress (i.e., disease susceptibility; see below) have been suggested to arise because of differences in genetic, biochemical, hormonal and social factors that contribute to different responding and coping strategies [29].

It is widely accepted that the incidence of various stress-related disorders is different in men versus women—men suffer disproportionately more often from coronary heart disease and substance abuse, while women are more susceptible to stress-related psychiatric disorders, such as generalized anxiety disorder, acute and post-traumatic stress disorder, and major depression [30]. In both men and women, gonadal hormones are implicated as having critical roles. For example, in men, exposure *in utero* and early in life to sex hormones has been proposed to underlie the higher risk for alcohol addiction later in life [31]. In women, the incidence of affective disorders, such as major depression, is twice as high as it is in men [32, 33]. As stress is a mitigating factor in such disorders, much work has focused on examining sex differences in HPA, in order to shed light on this relationship.

• • • •

CHAPTER 2

The Human Brain

2.1 BASIC PRINCIPLES OF NEURAL COMMUNICATION

In order to place some of the concepts we will talk about below within the proper framework, we will briefly describe the major components of the CNS and their functions.

2.1.1 The Action Potential and Release of Neurotransmitters

The CNS consists of the brain and spinal cord and the nerves that allow them to communicate. The brain is made up of cells that can broadly be classified as neurons and glia. Both types of cells serve important functions, and there are disorders of the brain that can be attributed to each. There are billions of neurons of over 200 different structural types; however, glial cells outnumber neurons 10:1. Glial cells are typically smaller and provide essential functions, such as removing toxins, manufacturing nutrients, and providing physical support. Neurons, on the other hand, receive, process and send messages throughout the CNS. Thus, the neuron is the computational unit of the brain, in terms of neural processing and cellular communication.

The basic anatomy of a neuron is illustrated in Figure 5. A neuron is either at rest ('not firing'), or it is stimulated ('firing'). In the stimulated state, a nerve impulse (or action potential) is triggered at the cell body (groups of which are called grey matter), and then travels down the axon to the axon terminals. Although movement of the message along the axon is an all-or-none event, the speed with which the message moves is influenced by the degree of insulation (myelination) covering the axon; however the intensity or relative strength of the message is preserved all the way to the end of the axon. Specialized glial cells make up the fatty insulating layer of myelin around most axons in the adult brain, and myelinated axons are often referred to as white matter. The nerve impulse moves very quickly down a myelinated axon, jumping across the sausage-like deposits of myelin to small unmyelinated sections to influence ion conductance across the membrane, and hence, transmission of the message. For a more detailed overview of this process, there is an animation publicly available in McGraw-Hill's Online Learning Center for *Human Anatomy*, by Michael McKinley and Valerie Dean O'Loughlin [reference], entitled "The Nerve Impulse," which you can access here: http://highered.mcgraw-hill.com/sites/0072495855/student_view0/chapter14/animation__the_nerve_impulse.html

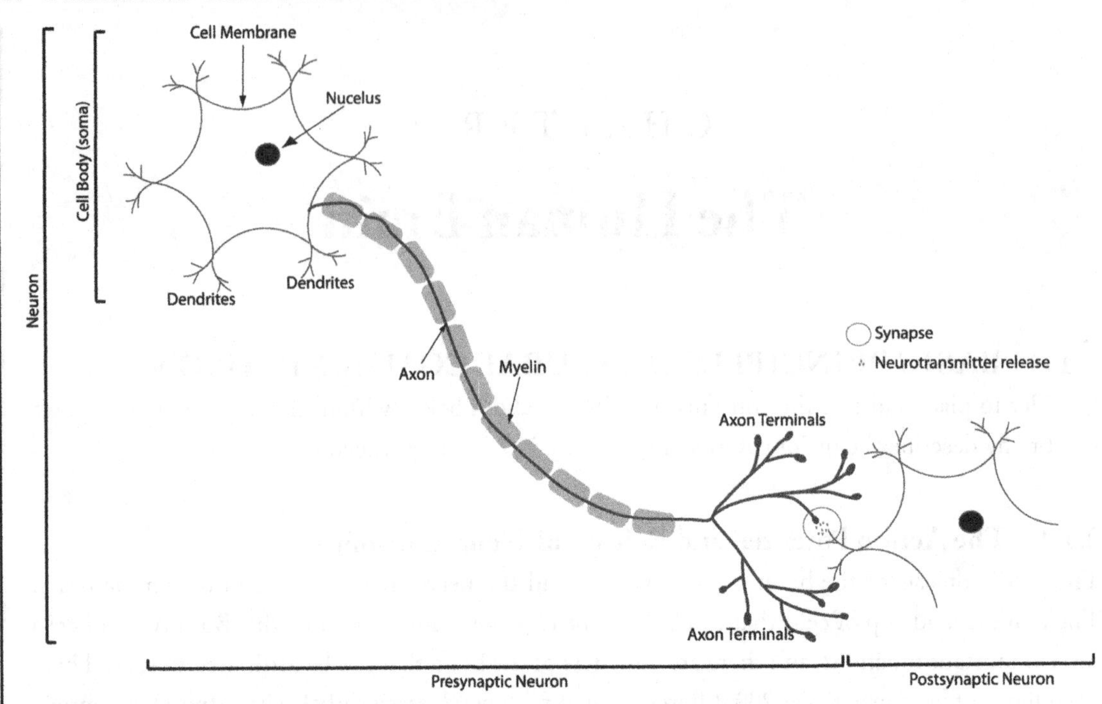

FIGURE 5: A simplified schematic of the major components of a neuron. The cell body (gray matter) contains structures called dendrites that receive messages from many other neurons. If the sum of these messages is sufficient to alter the voltage potential of the neuron when it is 'at rest', the neuron will fire (termed an action potential) and the message will be transmitted electrically down the axon to the axon terminals. An action potential is an all-or-none event once the voltage threshold is reached, and the strength or intensity (magnitude) does not change as the message travels down the axon. However, the speed at which it travels is dependent on the extent of fatty insulation (myelin; white matter) that covers the axon. Myelination of axon fibers increases the speed with which the action potential travels to the axon terminals. Once the electrical signal reaches the axon terminals, it causes changes there that result in the release of chemicals called neurotransmitters into the space (synapse) between the transmitting neuron (pre-synaptic neuron) and the receiving neuron (post-synaptic neuron). Neurotransmitters act on specific receptors (like a key fits into a lock) on dendrites or the cell body of the postsynaptic cell, altering its voltage potential. If the firing threshold of the postsynaptic neuron is reached, the process will begin again in that cell.

Once the message reaches the axon terminals, it is capable of influencing other structures, such as other neurons, glands or muscles. In the case of neuron-neuron communication, specialized chemicals called neurotransmitters are released into a small space called the synapse by the sending neuron (presynaptic neuron) and these go on to act on receptors on the dendrites of the receiving neuron (postsynaptic neuron). In a living organism, the postsynaptic neuron simultaneously receives signals from many different presynaptic neurons, some which increase its firing potential, and some that decrease it. The summation process occurs at the axon hillock, the junction between the cell body and axon, and if sufficient stimulation is received from all sources of input onto the neuron, an axon potential will be generated here and propagate to the axon terminals, resulting in release of neurotransmitter and transmission of the message.

2.1.2 Regional Specialization

Although the brain is made up of only two major types of cells, the location of those cells matters with respect to function. It is useful to remember that the brain is built from the inside out. The oldest parts of the brain, from an evolutionary perspective, are those that develop first and are located below the cortical surface. Such regions are specialized for basic physiological drives such as hunger and sex (ex. hypothalamus). In contrast, cortical regions of the brain that are involved in higher-level cognitive processing most distinguish humans from other animals in their developmental capacity and complexity.

• • • •

CHAPTER 3

Developmental Programming of the Stress Response

3.1 INTRODUCTION

An organisms' early life experience is absolutely critical for determining future stress responding. Having said that, not every person that experiences an adverse early environment becomes susceptible to stress later in life. It has become increasingly clear that there is a complex interplay of genetic susceptibility and environmental factors that decide where along the vulnerable-resilient continuum an individual ends up.

Much of the work examining how stress responding develops has been fueled by the acknowledgement that adult stress responding is highly variable and individual. Understanding this variability comes from an appreciation of the interactive nature of genes and environment in configuring the final phenotype (see Figure 1). The magnitude and duration of GC output is influenced by variables such as sex and age, but even within a seemingly homogenous group, there are individual differences. One factor that has emerged as being important for contributing to this variability is the early environment. Many aspects of the stress response are not fully developed at birth and in fact, some regulatory regions, such as the prefrontal cortex (PFC), are not fully developed until early adulthood [34]. By maintaining plasticity, the stress response can be molded by aspects of the environment to fine tune responses to the prevailing conditions. This is not to say that the prenatal environment is unimportant for programming stress responding, but for the purposes of this book, we will focus on the programming events that take place during postnatal life.

The idea of developmental programming—early experiences exerting important and long-lasting effects on adult phenotype—has been supported by numerous studies over the years using rodents, primates, and humans. Indeed, it has been proposed that early adversity such as child maltreatment increases the risk for developing depressive disorders, by sensitizing the neurobiological systems that are responsible for stress responding and adaptation [35]. There is a vast array of models presently in use in primates and rodents that are designed to mimic the kinds of early experiences of humans that have been associated with long-term outcomes. In general, these models invoke some form of stressor or adversity prenatally or postnatally, and the effects are then observed at a later developmental stage, usually adulthood, but also adolescence. As we will discuss in great detail in this book, adolescence itself is a critical period for developmental programming, making the situation even more complex.

3.2 VULNERABILITY VERSUS RESILIENCE

The importance of ones' experiences during early life for shaping adult responses cannot be overstated. Chronic negative experiences such as neglect, abuse, or severe trauma have been repeatedly associated with the development of mood disorders, such as depression, anxiety disorders, and substance abuse [36, 37, 38, 39, 40, 41, 42] although there exists large individual differences in the incidence of such 'stress-related' diseases (see Fig. 6). Early adversity in the form of child maltreatment impacts negatively on health, academic functioning, and even economic productivity [41]. Nonetheless, there is wide variability in the outcomes of such maltreatment, and at least some of that variability is due to genetic factors. Even with respect to the effects of early experiences on development of the HPA axis, the findings supporting abnormal development in children and

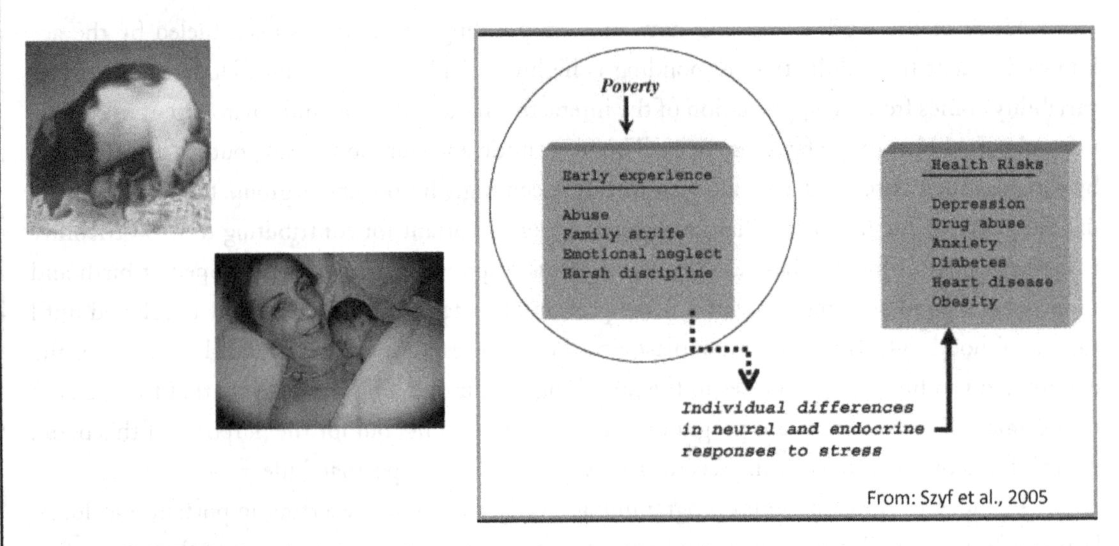

FIGURE 6: Adult mammals such as rats and humans exhibit individual differences in measures of stress responding. While some of this variability is the direct result of genetic differences among individuals, it also arises through experience-dependent modulation of early life development. Importantly, individual differences in stress responding are associated with differences in vulnerability to various forms of adult psychopathology. Some of the early life environmental factors that have been identified as playing a role in the development of a vulnerable individual are depicted in the left-hand box within the green panel, and subsequent health risks for the adult are depicted in the right-hand box.

adolescents as a result of maltreatment are not consistent. In children that were maltreated, most of the research that has been performed has measured basal levels of cortisol, and while some results indicate heightened cortisol, others have reported suppression [41]. Clearly, more work is needed.

The variability in outcomes following exposure to an adverse early environment exists on a continuum that actually includes resilience. For example, less severe forms of stress early in life appear to 'inoculate' some individuals against deleterious consequences of later stress. This has been described in numerous species, including humans. Early experiences that are challenging, but do not overwhelm the capability to cope, appear to produce 'hardy' individuals [43, 44]; the process of coping actually makes subsequent coping efforts more effective and creates the basis for regulating arousal and being resilient [45].

One brain region that appears to be important for developing resilience is the PFC. We will discuss this region repeatedly throughout this book because the PFC continues to develop in humans across adolescence and aspects of its function are therefore programmed during this developmental period. Interestingly, recent evidence indicates that enhancing medial PFC (mPFC) outputs has the capacity to convey resilience to future stress exposure [46]. In rats, resistance develops to the negative effects of inescapable shock stress with prior experience to an escapable version of the stress—blocking activation of the mPFC prevents this resistance [47].

3.3 ANIMAL MODELS

Models of stress response programming are varied across species, and the long-term effects they mediate are dependent on a wide array of variables, including obvious ones such as timing of the manipulation, but also less obvious ones such as sex of the individual. Much information has been gained from studying non-human primates, as they display many similar characteristics to humans; however, there is a plethora of research using rodents, and some of it will be summarized here.

The idea of resilience has its roots in rodent research. In rats, brief periods of separation of the offspring from the mother serve to enhance the capacity to handle stressors later in life. Termed 'early handling' and first described by Levine [48], the original experiments were designed to examine the negative aspects of such an early life manipulation. It was hypothesized that the intermittent stress of enduring the separations would result in emotional instability in the offspring in adolescence and adulthood. In contrast, the results demonstrated that the offspring were able to deal more effectively with novel situations in adulthood [49], due to a dampened emotional response in such situations. Other work looking at more discrete measures of stress responding confirmed that rats handled intermittently during the first week of life were less reactive to stress in adulthood [50]. Interestingly, these effects were assumed to be due to alterations at the level of central processing of the stressor, as many indices of HPA axis functioning, such as basal levels of cortisol and ACTH (see Fig. 3), and the ACTH response to exogenous stimulation, are not dissimilar in handled and non-handled animals. What does differ is that resting levels of PVN CRH are significantly higher in non-handled animals, suggesting a means by which stress responses are heightened [51].

The early work in rats demonstrating that intermittent periods of separation from the mother alter later stress responding prompted work in squirrel monkeys that has been very fruitful in elaborating on the idea of resilience [52]. The initial studies were begun with a similar hypothesis to Levine—that brief separations of offspring from the mother early in life would result in greater emotional instability in the offspring later in life, but similar to the work in rats, instead, a model of resilience was born. Monkeys exposed to intermittent separation during early life grew up to display less anxiety-like behaviors, lower levels of GC release following stress exposure, and more effective termination of GC release [53, 54, 55].

A dampened stress response can be as disadvantageous as a heightened one, and it could be argued that what is most adaptive is the ability to regulate one's emotional state to the point of being able to have a flexible response—one that is suited to the present threat. Thus, conceptualizing adaptive stress responding that results in a phenotype that is resilient to the detrimental effects of stress should conceivably include the capacity to maintain a variable coping style. In an interesting study, rats were divided into groups based on their initial response to being mildly restrained on two occasions [56]. Rats were distinguished based on their attempts to escape and were labeled active, passive, and variable copers. These three groups were then subjected to a period of chronic mild stress for 20 days, and their behavioral flexibility was assessed at the termination in the form of response to an acute stressor (forced swimming). The variable copers displayed more adaptive behavioral responses to the forced swim test and had a distinct cardiovascular and neural response to this stressor. Interestingly, a brain region called the bed nucleus of the stria terminalis (BNST) was more active in the variable copers than in the other two groups [56], suggesting this area may be involved in resilience. Although studies of the brain circuitry underlying resilience are still in their infancy, a role for the BNST would be very interesting, given that there is a direct connection from the BNST to the PVN to modulate responses to stressors [57].

Another exciting aspect of the possibility that the BNST plays a major role in resilience is the suggestion that it receives an indirect projection from the medial PFC [57, 58]. The PFC plays a prominent role in integrating responses to psychological stressors [59] and is important for behavioral flexibility [60], a trait that has been associated with positive long-term adjustment in humans [61]. The role of the PFC in decision-making [62], as well as in social interactions and emotional processing [63], make it a brain region particularly suited to influence flexible behavioural responses to stress, and it will be discussed throughout this book.

Maintaining a variable coping response necessitates being able to regulate one's emotional state. The idea that resilient individuals have an increased capacity for arousal regulation has been demonstrated best in squirrel monkeys. As mentioned above, the PFC is involved in emotional regulation. In humans, imaging studies have shown that larger volumes of PFC are associated with lower impulsivity [64], and, in squirrel monkeys, early life intermittent stress enhances the ability to inhibit a preferred response, a function also associated with the PFC [65]. Recent work in monkeys extends the human imaging findings by showing that expansion of the ventromedial PFC that occurs in monkeys exposed to mild early stress is consistent with increased white matter myelination in this area [45]. The increased myelination suggests an enhanced functional capability of this region of the cortex, one of the functions of which is to regulate arousal and resilience.

3.4 POTENTIAL ROLE OF MATERNAL CARE

The research findings summarized above no doubt leave the reader with the idea that there is something important not only about early life but about the interpersonal relationships that are formed during that time for the development of the stress response. In fact, it was proposed many years ago that the early handling effect in rodents was mediated by maternal care. The role of maternal care in programming the stress response has been reviewed many times before (see [66] for a recent review), and we will present only a critical summary here.

Newborn rat pups experience contact solely with their mother and siblings during the first two weeks of life, being completely dependent on the dam for survival. Thus, maternal care is a critical link between pups and the early environment [67, 68], and, given the highly variable nature of maternal behaviour in laboratory rats [69, 70, 71], varying levels of maternal care could play a role in programming individual differences in adult offspring stress responding. In laboratory rats, systems important for the adult expression of behavioural and neural responses to stress are programmed during the first two weeks of life [48, 72], and levels of specific maternal behaviours are associated with the development of these responses [67, 71, 73, 74]. Specifically, mother-offspring interactions are important for the development of adult behaviours, including learning [75], fear and anxiety-related behaviours [76], and the development of psychopathology [77]. More recent work has identified specific epigenetic changes that occur as a result of increased maternal care and that mediate the long-lasting changes in the offspring. In this model, increased levels of licking and grooming of pups by the rat dam leads to increased levels of a factor that modulates gene expression and in this case leads to increased expression of GR levels in the hippocampus [78]. Recall that binding of GCs to GR in hippocampus is involved with recovery following exposure to a stressor, and, therefore, increased levels of GR would be associated with a more effective termination of the stress response and potentially with more adaptive stress responding.

In order to place the apparent connection between variable maternal care and individual differences in offspring stress responding within a more ecologically valid framework, we manipulated levels of maternal behavior by exposing rat dams to a predation threat on the day of giving birth. Doing so increased maternal behavior in these dams for the next 5 days, which was accompanied by increased levels of estrogen receptors in a brain region critical for maternal behavior, and led finally to the epigenetic transmission of increased maternal behavior to the next generation [79]. Offspring of dams exposed to the predation threat (and increased maternal care) displayed sex-specific programming of anxiety-like behavior, but this was not accompanied by any alterations in GR transcription levels [80].

The work summarized above places the utmost importance on maternal care as the mediating factor in offspring stress response development, and this work has had large effects on the stress response development field. However, there is evidence to suggest that maternal care is neither necessary nor sufficient for offspring HPA development [81]. For example, exposure to novelty necessitated by early handling (EH) paradigms has long-term consequences on offspring, independent of the effects of maternal care [81]. An alternative view favors maternal care as a factor capable of modulating the direct effects of neonatal stimulation on long-term changes in stress responding.

3.5 IMPORTANCE OF PREFRONTAL CORTICAL AREAS

In addition to its role in higher order cognitive processes and emotion regulation, the PFC has been shown to significantly affect stress responding. Figure 7 depicts subdivisions of the PFC in human brain. In fact, stress induces activation of the PFC, as indicated by an increase in immediate early gene expression [82], and there is a high concentration of GR found in the medial portion of the PFC in rats [83] and primates [84]. In turn, PFC activation increases plasma GC levels (reviewed in [63]), and focused lesions result in enhanced cfos mRNA expression (indicating neural activation) in the rat PVN following restraint stress, but not ether stress [83]. As highlighted by this latter study, the role of the PFC is far from straightforward in rats. Some studies have noted very few effects of lesions to [57], or temporary inactivation of [85], the medial PFC on HPA axis responses to a stressor. Others, however, have noted involvement of the medial PFC that is dependent on the subregion examined [86] and the type of stressor [83, 87].

In healthy human volunteers, negative associations have been found to exist between activation of the medial PFC (as indicated by increased glucose metabolism) and salivary cortisol levels, suggesting a dampening role of this brain region on HPA axis output [88]. The medial PFC also sends projections to neurons in the sympathetic nervous system, which exert an inhibitory control over the release of hormones in response to stress-induced SAS activation [89]. However, the medial PFC is not homogenous in its control over the HPA axis. While dorsomedial parts are involved in inhibition of HPA axis responding, more ventral regions—in the primate, this corresponds to a region called the subgenual PFC—have been associated with HPA axis activation. In a study of rhesus monkeys, individual differences in HPA axis output across a range of non-threatening and threatening environments were consistently predicted by PFC activation [90]. In contrast to the effects observed using inactivation techniques (lesion or chemical), increasing the neural activity in

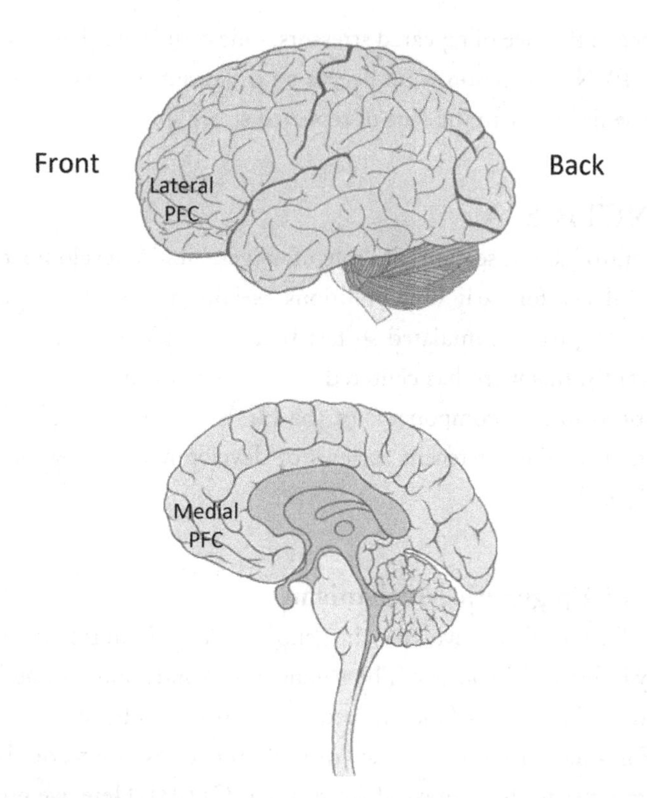

FIGURE 7: A schematic of the human brain, including a sagittal section in the lower image. The prefrontal cortex (PFC) resides at the extreme front, and different functions have been attributed to subregions that are more medially versus laterally situated.

the medial PFC partially suppresses activity of the HPA axis in response to restraint stress in rats [85]. Interestingly, in these same rats, increasing medial PFC activity suppressed activation of the hypothalamic PVN, suggesting an important role for medial PFC in HPA axis regulation [85]. Although there are no direct neuronal connections between the medial PFC and the PVN, there has been a projection identified from part of the medial PFC to the BNST, and the BNST neurons then project to the PVN, forming an indirect route [58].

Importantly for individual variability in stress responding, there is recent evidence that the PFC is involved in top-down control of habituation of the stress response. The ability to habituate, or display a lessened response to a stressor stimulus that is repeatedly encountered, is highly adaptive and is observed across species. In rats, the PFC appears to be critical for habituation of responses to repeated episodes of restraint stress. Recall that the BNST (see section 3.3) has been suggested to play a role in resilience in the face of repeated stressors. One could speculate that a high-functioning medial PFC-BNST-PVN connection would serve an organism well with respect to adaptive habituation and response flexibility in the face of repeated stressors.

3.6 EPIGENETICS

As indicated in the Introduction section, the role of epigenetics in developmental programming is monumental. The evidence for early life conditions exerting long-term epigenetic modifications to adult stress responding has accumulated so fast that it is difficult to imagine the field without this knowledge. Recently, this work has centered around the effects of adverse early life events on epigenetic modifications to core components of the HPA axis (for example, CRH release or GR levels), and there are a number of recent reviews of developmental programming via epigenetic mechanisms (ex. [91, 92]).

3.6.1 Overview of Epigenetic Mechanisms

There are a number of mechanisms involving the genome, but excluding changes to the genetic code itself, that ultimately influence phenotype. These generally include non-coding RNA species, methylation of DNA, and modifications of histone proteins (most notably, acetylation, phosphorylation, and methylation). For a more thorough review of these processes, the reader is referred to a paper that discusses the processes in the context of stress and GCs [93]. Here, we present an overview of the pertinent mechanisms for the reader to appreciate their importance to the work discussed. See Figure 8 for a depiction of the major mechanisms we will discuss.

Within the DNA sequence, methylation (the process of adding a methyl group) of cytosine residues that lie adjacent to guanine residues (aptly called CpG sites) represents one of the most well studied epigenetic marks. These CpG sites are often clustered together and located within the promoter regions of genes, the regions that attract the molecular machinery necessary for the gene to be expressed. An increase in methylation of CpG sites within a promoter region is associated with a reduction of expression of that gene, because methylation interferes with the binding of factors that lead to expression. Methylation of DNA is mediated by specific enzymes (DNA methyltransferases; DNMTs), of which there are a few varieties, some of which are involved in maintaining the methylation status of the DNA throughout the lifespan and others that are involved in *de novo* methylation.

The other set of epigenetic modifications that have been studied widely are modifications to the histone proteins that package DNA into chromatin. The more tightly that the DNA is wound around these histone proteins, the less likely gene expression will occur. Modifications such

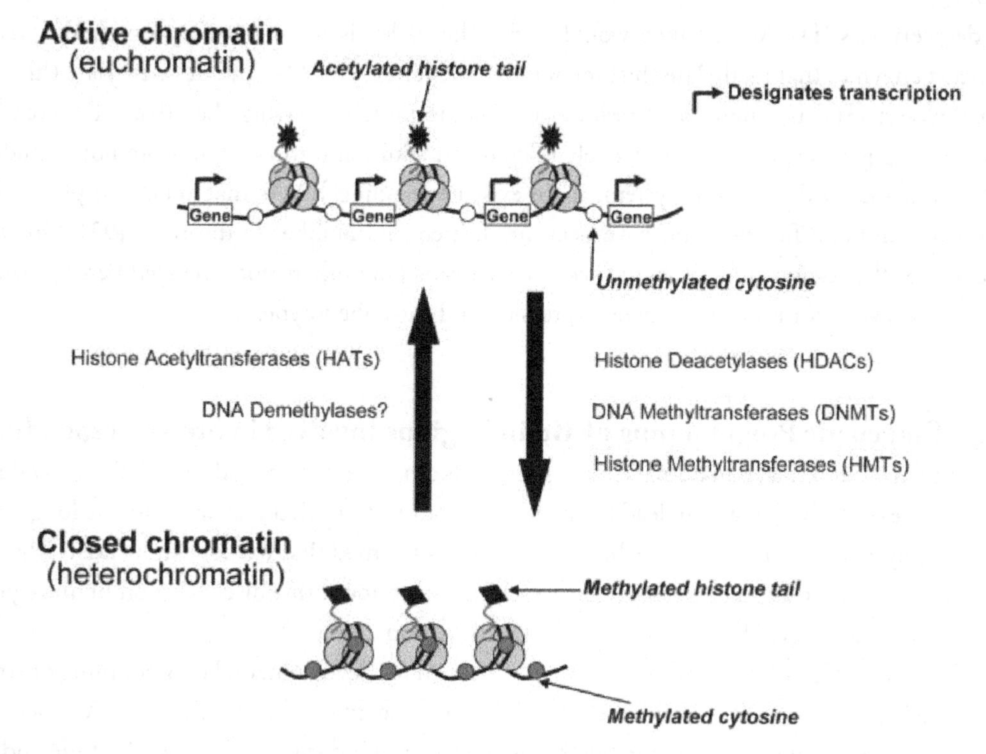

FIGURE 8: An overview of the major epigenetic modifications that occur in the brain. Genes (sequences of DNA) are organized as part of chromatin, which also contains proteins called histones. In order for genes to be expressed, the DNA sequence needs to be 'read' (a process called transcription) by molecules that require physical access to the sequence. When the chromatin is in a closed state, it is much more difficult for transcription to occur. There are various epigenetic modifications that can occur to the DNA and the histone proteins that loosen the chromatin, rendering it active, and increasing the likelihood of transcription and, thus, gene expression. Moving between closed and active chromatin states includes modifications such as adding methyl and acetyl groups to specific amino acids that make up the histone proteins or to specific base pairs that make up the DNA. Specific enzymes, such as those illustrated in this figure, are involved in mediating these modifications. From Figure 1 from *J Child Psychol Psychiatry*, 2011 Apr;52(4):398–408. Roth, TL & Sweatt, JD.

as methylation and acetylation of specific amino acids that make up the histone proteins dictate how tightly packed the DNA and proteins will be and, therefore, the extent of gene expression. For example, acetylation of amino acid residues of histone proteins is achieved by the action of a specific group of enzymes called histone acetyl-transferases (HATs) and is associated with a transcriptionally active state, resulting in more gene expression. Acetylation can be reversed by histone de-acetylases (HDACs), which would lead to lower levels of gene expression. Additionally, there exist enzymes that methylate (histone methyltransferases; HMTs) and de-methylate (histone de-methylases; HDMs) amino acid residues of histone proteins, having the effect of decreasing and increasing gene expression, respectively (Fig. 8). In addition to these more commonly studied epigenetic marks in the nervous system, there exist non-coding RNAs that appear to play a significant role in brain function and have been implicated in a number of disorders [93]. One can envision that the combination of all of these mechanisms (and others not discussed here) provides a powerful capacity for influencing gene expression and thus phenotype.

3.6.2 Epigenetic Programming of Brain Regions Involved in Stress Responding

Sufficed to say that prenatal maternal experiences, including stress, postnatal maternal care, and neglect or abuse of the infant, all can lead to epigenetic modifications that may underlie the long-term effects on physiological, neural and behavioral responses to stress that we now know exist. It is not the intent to review this material again here, but instead to focus this discussion on brain regions that provide input to the HPA axis and modulate stress responding.

A role for epigenetic modifications as a mechanism for developmental programming of stress responding began with work showing the role of maternal care in rats (see Section 3.4). As discussed there, pups reared by mothers that provide higher levels of care during the first week of life end up with higher levels of hippocampal GR in adulthood, which partly explains their more efficient stress response. It turns out that increased licking and grooming of pups by the dam increases levels of nerve growth factor-inducible protein a (NGFI-A), a factor that can modulate gene expression. In this case, NGFI-A interferes with methylation within a particular promoter region of the hippocampal GR gene, having the effect of increasing expression of GR protein in this region [94]. Recall that the hippocampus plays a large role in regulating negative feedback of the HPA axis and, thus, increasing levels of GR protein is consistent with a more efficient feedback mechanism.

Evidence for long-term changes to the PFC that are associated with childhood maltreatment is mixed, with clearer evidence for structural alterations than functional ones [95]. Nonetheless, epigenetic dysregulation in PFC and hippocampus has been associated with the development of mental illness [96]; thus, such processes could underlie the long-term effects on cognition and emotional stability that accompany early life adversity. The fact that childhood abuse and neglect results in increased susceptibility to later cognitive deficits and various psychiatric disorders has led to the hypothesis that neural plasticity within this brain region is likely compromised. Evidence to support such a claim has arisen from studies in which the PFC of adult rats was examined following exposure to maltreatment in early life. Persistent changes in DNA methylation within a gene region that encodes for a protein involved in various forms of plasticity in developing and adult brain, brain-derived neurotrophic factor (BDNF), was observed [97]. These alterations were passed on to the female offspring of the affected adults [97].

More recently, work has begun to delineate the long-term effects on the PFC of stress experienced during the adolescent developmental period. Development during this period is very important for the refinement of cognitive abilities; thus, it is likely that stressors experienced in the environment at this time will shape the cognitive abilities of the emergent adult phenotype, such

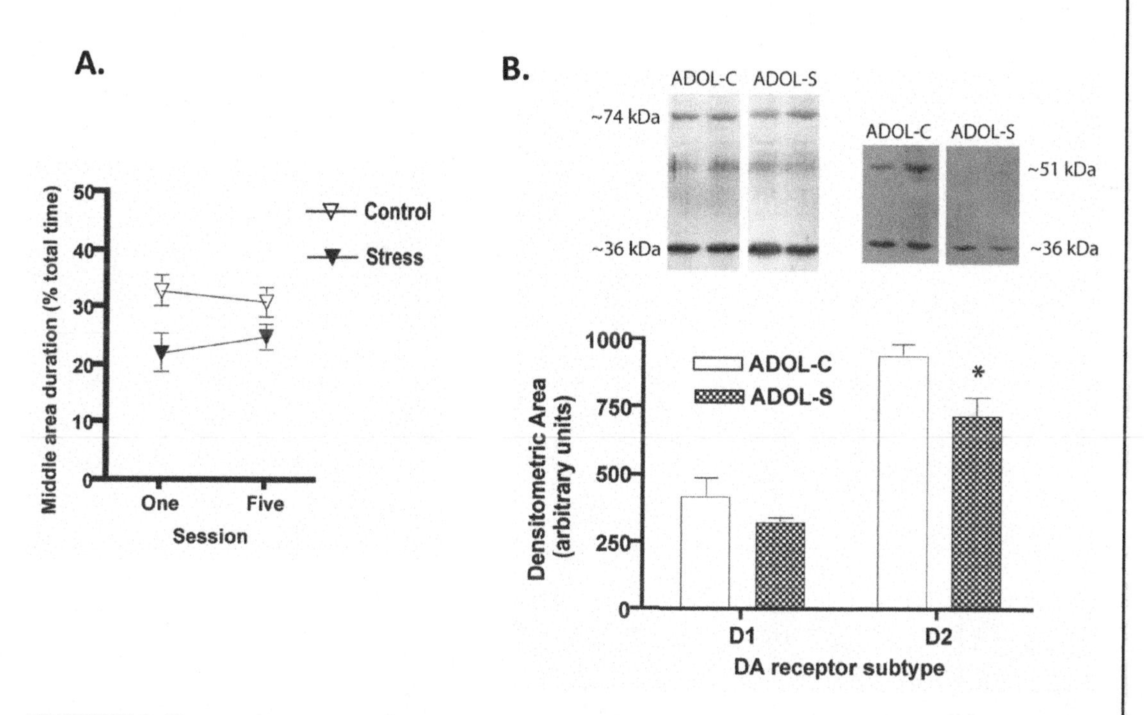

FIGURE 9: Repeated exposure of rats to stress across the adolescent period results in altered levels of the D2 dopamine receptor in the prefrontal cortex in adulthood. (A) Adolescent rats exposed repeatedly to cat odor stimuli (Stress) show defensive behavioral responses during the exposure sessions, such as reduced time spent in the middle of the exposure arena (data taken from [20]) relative to Controls. (B) In adulthood, those rats that had been exposed to cat odor during adolescence (ADOL-S) had significantly lower levels of the D2 (but not the D1) dopamine receptor relative to those that had been exposed to a control odor during adolescence (ADOL-C). Since prefrontal dopamine transmission is important for cortical modulation of HPA activity in the adult, these findings suggest an experience-dependent alteration of prefrontal function related to stress responding.

as the ability to exert inhibitory control over stress responding. In our lab, we exposed developing rats to a potent, species-typical stressor stimulus (predator odours) on several occasions during the adolescent period and subsequently observed altered levels of the D2 dopamine receptor in the PFC of these rats in adulthood, relative to control rats (see Fig. 9). These findings support the idea that adolescent stress will shape development of the PFC. The functional significance of the observed changes is still not well understood but is actively being investigated.

• • • •

CHAPTER 4
The Importance of Adolescence

4.1 INTRODUCTION

Traditionally, most research investigating the impact of stress on brain development and adult behaviour has focused on early life and childhood, in both animal models and human studies (ex. [98, 99]). However, mammals, and humans in particular, are afforded a protracted period of brain development that extends beyond childhood. The period between childhood and adulthood is termed 'adolescence,' a word derived from the Latin word 'adolescere,' which means 'grow up.' Presently, there is a surge of interest and research attention devoted to understanding the typical course of adolescent brain development. It is now recognized that adolescence is a special developmental period, in that the changes that take place are not simply a continuation or extension of childhood development that come to completion as the individual reaches adulthood. Rather, there is a series of developmental processes that are separate and distinct from the changes that occur in childhood, and these are signaled for initiation during or after puberty. Some of these changes are signaled by the rise in circulating gonadal hormone levels that occurs during puberty. Thus, puberty is often thought of as the beginning of adolescence. Other changes appear to be hormone-independent and don't coincide with puberty but instead emerge later on during the adolescent period. Here we will outline some of these changes and discuss how they may be regulated by the prevailing demands of the environment (such as whether or not there are stressors present), and also how they impact adult stress responding.

4.2 OVERVIEW OF PUBERTAL/ADOLESCENT PHYSICAL DEVELOPMENT

Puberty itself designates the attainment of sexual maturity and usually denotes the beginning of the adolescent period. Developmental changes continue to take place throughout the brain for about a decade longer in humans, particularly in regions that are important for cognition, such as the PFC. Thus, humans and other mammals become capable of reproduction well before they have reached full cognitive maturity.

Typically, girls observe their first period around age twelve [100, 101, 102] and boys begin to ejaculate around age thirteen [103], although for both measures there is a wide margin of variability around the mean. Those who show these landmark signals of puberty more than a year or two earlier or later than the average are considered to be precocious (early) or delayed maturers, respectively (ex. [102]). Currently, girls are considered to mature more quickly than boys, on average. Although girls do show hallmark features of puberty at an earlier chronological age than boys, the fact that these hallmarks are very different for boys vs. girls (e.g., ejaculation vs. menstruation) makes it challenging to directly compare boys and girls, in terms of degree of maturity, especially in regards to brain development. For example, girls begin menstruating on average one year sooner than boys begin ejaculating, but menstrual cycling may be irregular for a period of several years during adolescence, and ovulation, which is necessary for fertilization, may not occur during each cycle. On the other hand, full adult fertility may be achieved sooner after the first ejaculation event in boys. Again, there is a great deal of variability among individuals in the progression of pubertal changes, in addition to the variability in puberty onset.

Early during puberty both boys and girls undergo a growth spurt—a rapid increase in height and weight—and this is accompanied by changes in skeletal structure and proportion of body fat, which altogether result in boys' and girls' bodies gaining the characteristic physiques of adult men and women. Prior to puberty, girls are on average a little bit taller than boys [104] and have slightly less waist fat but more hip fat than boys [105]. At the completion of puberty; however, the average man is several inches taller than the average woman [104], and women have almost 50% less waist fat than men (but still more hip fat; [105]). The growth spurt, and the other physical changes associated with puberty that occur in the body's periphery (outside the CNS), are directed primarily by increases in levels of testosterone and estradiol secreted by the gonads, which are then distributed throughout the body to act on receptors localized in various organs and tissues. While testosterone plays a particularly prominent role in male differentiation, and estradiol, the major human estrogen, is necessary for many aspects of female differentiation, both testosterone and estradiol are produced in both men and women, and both steroid hormones play important roles in pubertal development for each sex. For example, the growth spurt during early puberty is initiated by an increase in circulating estradiol levels in both sexes [104].

Changes in pubertal hormone levels actually begin taking place a couple years before the hallmark features of puberty start to emerge [104]. The hypothalamic-pituitary-gonadal (HPG) axis is considered to be the body's major system for regulating patterns of secretion of sex steroids and plays a central role in pubertal development. This system is comprised of the hypothalamus and pituitary gland within the brain, and the gonads—testes in males and ovaries in females—in the periphery. In brief, neurons within the hypothalamus produce and secrete gonadotropin-releasing hormone (GnRH), which acts on the anterior portion of the pituitary gland to cause the production of luteinizing hormone (LH) and follicle-stimulating hormone (FSH). These hormones are secreted into the bloodstream and act on the gonads to induce production of sex steroids, culminating finally in the characteristic increases in testosterone, estradiol, and other steroid hormone levels observed at puberty. These steroid hormones are all derived from a cholesterol base and are produced and released by cells of the endocrine glands of the HPG axis, among other areas. Once released, the hormones can either act locally in the surrounding tissue, or they can be secreted into the bloodstream and then taken back up into organs and tissues at other locations in the body to act on receptors there. There are a variety of receptor subtypes that are able to bind these steroid hormones, and the effect of the hormone depends on the receptor type to which it binds. Thus, the body is able to simultaneously regulate different actions of these hormones at different locations, by localizing the receptor subtypes in a region-specific manner. Also the HPG axis is self-regulatory, in that release of gonadal hormones results in a negative feedback loop at the level of the hypothalamus, by binding to receptors that shut off further release of GnRH [106].

Measuring levels of hormones, per se, is informative but doesn't provide the full picture. A hormone must interact with a receptor (similar to a key interacting with a lock) and this interaction has physiological effects (like the opening of a door when a key is turned in a lock); thus, the nature of the interaction is of utmost importance, since a small change to that interaction could potentially cause a very different physiological effect (like the popping of a trunk when the key is turned in a different lock). As an example of how different receptor subtypes for a single steroid hormone can mediate different actions, we will consider estrogen receptors. There are two classes of estrogen receptor that recognize estradiol; one class is an intracellular receptor (i.e., estrogen receptor alpha or beta (ERα or ERβ)) that resides inside the cell. If estradiol becomes available to bind to the receptor its overall conformation will change, such that, when estradiol-bound, there is movement of the hormone-receptor from the cytoplasm of the cell to the nucleus, where the complex can then bind to specific regions of DNA, such as promoter sequences, and influence expression of genes associated with those sequences [107, 108]. If estradiol becomes dissociated from the receptor inside the nucleus, the empty estrogen receptor can also be recognized by particular segments of DNA and bind to it to alter gene expression [109].

Estradiol and other steroid hormones have high lipid solubility and thus can easily traverse cellular membranes, which contain many phospholipids. Thus, an increase in circulating estradiol in the bloodstream during puberty results in increased absorption from the vasculature into surrounding tissues, and the estradiol eventually passes into cellular nuclei, where it alters the intracellular estrogen receptors' functional capacity to act as a transcription factor. In this way, estradiol secretion can regulate gene expression for a number of different genes that have promoter sequences recognizing the estrogen receptor. It is by this mechanism that rising levels of estradiol induce many of the physical changes associated with pubertal development.

The second class of estrogen receptor in existence in key endocrine regions of the brain is a G protein-coupled receptor (i.e., G protein-coupled estrogen receptor 1 (GPER; also known as GPR30); [110]). G protein-coupled receptors are membrane embedded and can have roles in membrane production, protein synthesis, and steroid production. Upon binding estrogen, a cascade of intracellular signaling events takes place. This involves increases in intracellular calcium levels, which, among other functions, can increase neurotransmitter release into the synapse. GPER activation also initiates downstream activation of protein kinase molecules that mediate a variety of cell signaling events. Overall, the nongenomic cell signaling effects of GPER activation by estrogen occur over a short timeframe, relative to the changes in gene expression induced by ERα or ERβ activation, and thus the rapid GPER-induced molecular signaling events may underlie short-term behavioural responses that are mediated by estrogen, such as stress responding in the presence of a threat stimulus. ERα and ERβ, on the other hand, appear to play a more direct role in regulating the estradiol-induced physical changes that take place during puberty in both boys and girls.

Thus, the HPG axis instigates a series of physical alterations during pubertal development in both sexes. This is accomplished through pulses of GnRH release into the anterior pituitary, which results in elevations in circulating levels of LH and FSH, followed by the production of gonadal hormones, namely androgens and estrogens. Upon secretion into the bloodstream, the hormone molecules can be distributed rapidly throughout the body to act on a variety of receptor types in different organs and tissues. Sometimes the hormone molecules may be converted by an enzyme into another form before being recognized by a specific receptor type and initiating a certain effect. For example, testosterone is the major androgen involved in the initiation and direction of male pubertal changes; however, testosterone can be converted to estradiol by the enzyme aromatase. The presence of aromatase in males results in the production of low levels of estradiol. Thus, estradiol is an active metabolite of testosterone, and, once it is made from testosterone, it will interact with estrogen receptors instead of androgen receptors.

These hormone/receptor interactions are responsible for guiding a typical pattern of effects in each sex. Over time, these effects manifest as the physical alterations that differentiate the male body from the female body. The typical progression of physical alterations at puberty starts with the growth spurt induced by estradiol. This occurs earlier in girls than in boys, as girls show an earlier rise in estradiol levels, as well as a greater peak in the levels observed [104]. Boys, on the other hand, show elevated estradiol levels a little later on, as testosterone levels rise first and then some testosterone gets converted to estradiol by aromatase, resulting in a slower onset of the growth spurt. However, the overall spurt is protracted in males, relative to females, allowing for the eventual greater body size in men [104]. Simultaneously, rising levels of androgens change the fatty acid composition of bodily perspiration in both boys and girls and also increase the production of sebum (oil) from the skin [111, 112]. These androgenic effects result in a change in body odour and often the appearance of acne, which normally resolves by adulthood. Also, following the onset of the growth spurt, a strength spurt occurs in boys. During this testosterone-driven change, males increase lean muscle mass, relative to females, whereas females gain an overall higher relative proportion of body fat [113]. The rate of muscle growth peaks in boys approximately one year after the peak in overall growth (height/weight). Furthermore, bone growth also accelerates during this stage, a process regulated by estradiol [114, 115, 116] that results in the characteristic differences in skeletal shape between adult men and women, such as wider shoulder blades in men and wider hips in women.

Aside from these gross changes in body size, shape, and constitution, the major physical changes associated with puberty are the development of the gonads (gonadarche), and the appearance of secondary sex characteristics, which refer to features that are not directly involved in reproduction.

The male gonads, or testes, begin to enlarge at the onset of puberty. Testicular maturation and the appearance of male secondary sex characteristics are directed by increases in levels of androgens and by their interactions with their receptors [117, 118]. Prior to puberty, the average testicular volume is only about one or two ml; however, the testicles continue to grow for about six years after the onset of puberty, eventually reaching a maximal adult volume of approximately 20 ml. After about one year of testicular growth, the penis also begins to enlarge, and the opening of the foreskin widens progressively, eventually allowing for retraction behind the glans or head of the penis. The testicles are housed in a protuberance of skin and muscle tissue called the scrotum, which also grows at puberty to accommodate the growing testicles and begins to hang lower, allowing the testicles to maintain a cooler internal temperature, which is required for sperm fertility. Sperm production begins in the testes shortly after puberty onset, and most boys become potentially capable of fertilization around age thirteen, although the first ejaculate usually lacks sperm [119, 120, 121, 122]. It is thought to take one to three years for boys to gain their full fertility potential.

Growth of the female gonads, the ovaries, is more obscure, due to their deep internal location. More obvious, outward signs of puberty in females are thelarche—the onset of breast development, and menarche—the appearance of the first menstrual bleed. Ovaries, like testes, produce testosterone, most of which is rapidly converted to estradiol by aromatase prior to secretion into the blood. Levels of androgens and estradiol, as well as progesterone and prolactin, are all important in regulating female pubertal changes [123, 124].

Thelarche usually occurs one to two years before menarche. Often, one or both areolae will pop out first. By the time menarche occurs, the growth has usually extended to the area around the areolae. The breasts continue to enlarge for at least another year, and the secondary mound consisting of the areolae and nipples eventually disappears into the larger surrounding breast mound. For about two years after thelarche, the ovaries, uterus, and follicles inside the ovaries continue to grow [125]. Also, in response to rising levels of estradiol, the vaginal mucosa becomes thicker and a duller pink in colour and may discharge a whitish substance called leucorrhea [125, 126].

In the adult female, menses will occur on average every twenty-eight days and usually lasts between three to seven days each cycle. Cycling can often be irregular and anovulatory during the first couple of years, though [127], and the menstrual flow tends to be heavier and last longer than in the adult.

In boys and girls both, pubarche, the appearance of the first pubic hair, is often the second outward sign of puberty and is initiated by rising levels of androgens. It usually occurs shortly after the testes begin to enlarge in males and after thelarche in females, although it can occasionally occur prior to thelarche [128]. The pubic hair first begins to sprout along the base of the penis in males and along the labia in females and over the next couple years gradually begins to fill the pubic triangle. Sometimes the pubic hairs will spread as far as the upper thighs and lower abdomen. In males, other areas of the body also begin to grow hair following pubarche, such as the underarms, the perianal region, the upper lip, the sideburns, the areolas, the beard area, and eventually the arms, legs, chest, abdomen and back. Girls also grow underarm hair and may show hair growth in some of the other areas as well. There is a wide range of normal patterns of hair growth in both boys and girls.

Rising levels of androgens also cause the voice box or larynx to grow, which in males is prominent enough to produce a characteristic deepening of the voice and a visible 'Adam's apple' in the throat. The change in voice pitch is usually complete within a couple years of the first ejaculation.

As mentioned, there is a lot of individual variation in the timing and progression of the physical changes associated with puberty in both boys and girls [125, 126]. The most widely accepted method of classifying individuals into different stages of puberty is the Tanner Scale, whereby the following categories are divided into a series of five stages: the amount of pubic hair that is present (in both males and females), testicular volume (in males), and breast development (in females; [125]). Although the typical course of hormonal and physical changes associated with puberty is well studied, the factors involved in the timing of puberty onset are not fully delineated. Specifically, it is unknown what causes the initial hypothalamic pulsing of GnRH release. One recent theory proposes that leptin levels, which rise at puberty, are involved in the initiation of the rise in GnRH levels. This theory is supported by the observations that the hypothalamus, the site where the pulsatile release of GnRH occurs, contains leptin receptors and leptin deficiency results in failure to initiate puberty [129]. Interestingly, in the years preceding puberty, hypothalamic GnRH pulsing only happens during sleep, but by the end of puberty, this circadian difference disappears [130].

4.3 THE ADOLESCENT TRANSITION

While puberty designates the attainment of sexual maturity—specifically, the capacity to reproduce—adolescence by contrast denotes the period of transition from childhood to adulthood and involves a variety of physiological and behavioural changes that appear to be 'plastic' to some degree. That is, the context in which the adolescent developmental transition takes place plays an important role in determining the adult phenotype. Adolescence therefore encompasses puberty, as well as a number of other changes that allow the individual to gain independence from parental or other childhood caregivers and perform adult tasks and duties, such as working and raising a family. Traditionally, adolescence has been thought of as involving a wider perspective than puberty. Changes in behaviour are often considered to be a major aspect of adolescent development and are clearly influenced by biological, environmental, and societal factors, including cultural norms, whereas changes in behaviour are rarely considered with respect to pubertal development. Puberty, rather, is generally thought of as involving mainly biological/physiological changes that prepare the body for reproduction. This has led to a historical division in the perspective that is generally taken when considering either pubertal or adolescent development, with puberty typically being considered as a series of biological changes and adolescence being considered as a time of rapid and profound behavioural change. However, we now know that environmental and societal factors, in addition to biological factors, can influence both the timing and progression of puberty, as well as influencing aspects of adolescent development and adolescent behaviours. Furthermore, upon reflection it becomes obvious that behavioural development is also critically important for puberty, since reproduction necessitates the expression of sexual behaviour. On the flip side of the coin, the behavioural changes associated with adolescent development must also involve structural and/or physiological changes in the brain, since we now understand that the brain regulates behavioural output.

The outdated view that puberty involves biological changes and adolescence involves behavioural or psychological changes must therefore be revised to incorporate a deeper understanding of the intimate ways in which biology and behaviour are tied together. Toward that end, in this chapter we will review the information that is currently available on neural changes that take place during adolescent development and examine how these changes relate to the emergence of characteristic adolescent behavioural changes, such as enhanced performance on complex cognitive tasks, increased inhibition and ability to delay rewards, increased risk-taking behaviours, and a shift in social focus from the family unit to friends and other peers. In particular, we will consider how stress responding changes over the adolescent period and into adulthood.

Note that the characteristic pattern of adolescent behavioural changes is relatively well conserved across mammalian species, indicating deep evolutionary roots and common bases in the structure and/or functioning of the nervous system. Some of the changes, such as increases in cognitive capability, are most prominent in humans, because the part of the brain most directly responsible for these abilities, the neocortex, is greatly expanded in humans, relative to other mammals. However, even in this brain region, many of the developmental changes that occur at the cellular and molecular levels in animal models can be applied to human cortical development. The overall greater cognitive capacity in humans relative to other mammals appears to be due mainly to the greater extent of cortical tissue in the human species. For example, the entire flattened neocortex is 3–4 mm thick and 24000 cm^2 in adult humans but only 100 cm^2 in the adult cat [131]. As it develops, the human cortex becomes much more convoluted into folds of sulci and gyri, relative to other mammals, but the folding pattern itself doesn't seem to play a significant functional role, other than allowing for more tissue to fit into the space [131]. Another aspect of human cortex that might be particularly relevant for its unique set of abilities is its laminar structure. The mammalian cortex develops into six distinct layers with different cellular and functional characteristics. In humans, the layers show different developmental trajectories, with the deeper layers completing development first and the most superficial layer exhibiting the longest phase of postnatal development [131]. In primate models, such as the macaque, however, this difference in timing of development among the layers has not been observed [131].

The particularly well-developed neocortex of the adult human allows for humans to be 'cognitive specialists,' in terms of our ethological survival strategy. In other words, we have evolutionarily capitalized on the strategy of expanding the neocortex, in order to compete with other species, as well as other members of our own species, for space and resources. Enhancing cognitive capabilities allows us to do this by 'outwitting' the others, in terms of space and resource allocation. This is not to say that we have all been consciously greedy in utilizing space and resources, since humans clearly have the capacity to empathize and regularly display altruistic behaviours toward nonrelatives and other species in kind. Rather, it is meant in the sense that our cognitive abilities—our abilities to plan and problem-solve and think abstractly—allow us to set up mental representations of an entirely different reality than what exists in the external world. We are then able to set about constructing a new reality in the outer world, based on our mental presentation, such as the construction of a city in a location where nothing but forest existed before. Our cortically mediated foresight thus affords us the ability to mastermind our surroundings and manipulate them at will, although it is not refined enough to predict all of the long-term consequences of the changes.

Some of the behavioral changes that take place during adolescent development are subserved by subcortical brain regions, such as the changes in social focus that come about as the 'social module' of the mammalian brain matures during adolescence [131]. These behaviours will still ultimately be influenced by cognition in adult humans, since the human neocortex develops direct or indirect connections with every other brain region, as well as an extensive network of interconnections among its own subregions. Some of these connections develop during adolescence, which has important implications for adolescent changes in stress responding. For example, as the PFC continues to be refined during adolescence and gains the capacity to exert inhibitory control over behaviours mediated by subcortical structures, it becomes more prominent in regulating HPA activity. This allows adult humans to cognitively fine-tune stress responding in real time, a process that can go awry in post-traumatic stress disorder [132].

A final point that should be reviewed before we outline what is known on adolescent brain development is that the neural processes taking place are dynamically regulated by the environmental context. This is a marriage of the two older, dichotomized viewpoints that held that development (both during childhood and during the less focused-on adolescent phase) was a result of either: 1) the expression of genes that are inherited by the individual at birth, or 2) exposure to particular environmental contexts during the developmental period (i.e., the supposed "blank slate" theory). We now understand that these two extreme viewpoints must come together in a middle ground theory (see Introduction and Figure 1). As we have gained a more and more advanced comprehension of the biological mechanisms underlying gene expression, it has become abundantly clear that which genes get expressed and how much expression occurs is dependent on surrounding environmental factors, not only during development, but on an ongoing basis. This implies that developmental events that take place during adolescence are sensitive to features of the environment, producing an adolescent critical period of development and a theoretical divergence point with respect to possible adult phenotypic outcomes (see Fig. 1), supposing that two genetically identical individuals are exposed to different environmental conditions. It is proposed that this developmental flexibility has evolved in order to optimize fitness over the lifespan. Fitness in this sense refers to the ability to survive and reproduce. We will revisit this theme in subsequent sections and continue to discuss some of the epigenetic mechanisms that allow gene expression to differ among individuals based on factors other than differences in the genetic sequence (see Section 3.6.1), with the idea that these processes optimize the adult phenotype for survival and reproduction in the context of the prevailing environment. Note that this theory implies that if the features of an individual's environment were to change dramatically between adolescence and adulthood, the individual could be permanently maladapted to its adult environment.

In the following section, we will review adolescent brain development, with a particular focus on brain regions that are known to exert considerable control over the HPA axis in adulthood—namely, the hippocampus and cerebral cortex.

4.4 ADOLESCENT STRESS RESPONDING AND ONGOING BRAIN DEVELOPMENT

Upon observation, adolescents appear to possess unique stress response characteristics. This has been confirmed in both human and animal studies, which demonstrate that adolescents show differences relative to adults in HPA activity and behaviours following exposure to a stressor; for example, a prolonged elevation in circulating GC levels [133, 134]. Such differences in stress responding between adolescents and adults (and also differences between adolescents and children) suggest an underlying basis in the normal series of developmental events that occur. This has prompted research investigating the normal pattern of adolescent brain changes, in hopes that a description of the typical pattern of adolescent neural development will shed light on the reasons for the unique features of adolescent stress responding. Here, we will review the overall pattern of human adolescent brain development and examine in relative detail particular developmental events that impact stress responding.

The developmental events that take place in the brain during adolescence are relevant not only for understanding adolescent stress responding, but also their flexibility provides the basis for plasticity in many aspects of the adult phenotype, such that the emergent phenotype will be tailored to meet the current demands of the environment.

4.4.1 Structural Changes

The most compelling data that have accumulated on human adolescent brain development have been derived from neuroimaging studies. Magnetic resonance imaging (MRI), in particular, has provided a detailed picture of the anatomical changes that take place in stress-related brain regions across the adolescent period at a spatial resolution never before achieved using other neuroimaging techniques (~1–3 mm; [131]).

Using this technique, researchers can scan participants' brains at defined intervals across development to track overall growth. The contrast between grey matter content and white matter content in the brain can also be visualized. Grey matter as revealed on an MRI scan is made up mostly of neuronal cell bodies and dendritic processes, as well as glial cells. Since the neuron is the computational unit of the brain (see Section 2.1), the grey matter represents the volume of brain tissue involved most directly in conducting neuronal computations. White matter content, on the other hand, is composed mainly of bundles of myelinated axon fibers and is thus very important for the transmission of information from neurons in one brain region to neurons in another brain region. Changes in white matter content may therefore be indicative of changes in myelination and connectivity or 'wiring' among brain regions that form functional neural circuits. Altogether then, quantification of grey matter and white matter in the whole brain and also in various defined brain structures and regions at different points across development provides an excellent measure of brain growth and structural change as development progresses.

In regard to human adolescents, MRI studies have shown that the total brain volume remains fairly stable across the adolescent period in males and actually declines slightly in females, with boys showing a peak total volume at approximately 14 years old and girls at only ten years old [135]. In order to put these findings for the adolescent brain into context, we must first consider the pattern of brain volume change leading up to adolescence and also what happens as the brain matures into adulthood. Based on the longitudinal data that is available so far, the general consensus is that brain volume changes in typically developing individuals follow an inverted-U function across the lifespan, with a gradual increase during childhood and a slow decline into adulthood, after the peak volume has been reached. The total number of neurons in the brain actually changes relatively little after birth. The vast majority of nerve cell division takes place prenatally, excluding adult neurogenesis processes that have recently been shown to occur in specific brain regions, such as within the dentate gyrus of the hippocampus [136] and the cerebellum [137]. Although these findings likely represent an important mechanism for adult plasticity of brain function, most of the neurons that will compose the adult human brain are already present at seven months *in utero* [131]. Changes in grey matter volume that arise postnatally are thus attributed mainly to changes in neuronal architecture and not changes in the overall number of neuronal cell bodies.

In early life, these changes in neuronal architecture may involve neuronal migration processes. Neurons are not generally produced at the same sites in which they eventually reside after they've matured. Human cortical neurons, for example, are born within the subventricular zone and then actively move to their final destinations within the cortex, giving rise to an 'inside-out' pattern of development, as the deepest layers mature first, and the youngest neurons move past older ones to get to the most superficial layer [131]. Most of this neuron trafficking occurs prenatally; however, some late-developing brain regions, such as the most superficial layers of cortex, continue to form in the first few years after birth. The changes in grey matter volume also represent changes in the number of dendritic processes, with a characteristic increase in the number of processes and the degree of branching of the dendrites throughout later childhood and early adolescence, a finding derived from post-mortem studies that has been confirmed in animal models [131]. There is also a corresponding increase in the density of synaptic contacts observed in human cortex, followed by a decline in this measure [131]. The decline is termed 'synapse pruning' and is thought to reflect a developmental refinement of the neural circuitry, as functional networks becomes more efficient at communicating information. The role of experience in synaptic pruning processes is poorly understood; however, in accordance with Donald O. Hebb's proposition that "neurons that fire together wire together," it is commonly supposed that excess synapses are lost as neuronal circuits strengthen their most active connections during adolescent development, and the choice of which synapses are lost and which become an integral part of the circuitry is therefore dependent on neural activity, which itself is dependent on input from the environment. Consider the following quote from Hebb's book, *The Organization of Behavior:*

> "When an axon of cell A is near enough to excite cell B and repeatedly or persistently takes part in firing it, some growth process or metabolic change takes place in one or both cells such that A's efficiency, as one of the cells firing B, is increased" [138].

The "growth process or metabolic change" to which Hebb refers could involve the loss of competing synaptic connections that would also influence the firing pattern of cell B, such as hypothetical inputs from cells C, D, and E. The loss of these connections would mean that the voltage potential of cell B would be more heavily impacted by its input from cell A, giving cell A a bigger role in determining the moment-to-moment state of cell B (i.e., resting or firing). If the firing of cells A, C, D, and E is dependent on environmental factors, this could provide a mechanism by which the preferred neuronal connections are selected and come to form the adult circuitry.

Functional neural circuitry of course involves both groups of neuronal cell bodies, as well as bundles of axons (white matter) to transmit information among the various brain regions interconnected by the circuit, and, interestingly, whole brain white matter volume steadily increases across development [135, 139], instead of showing the inverted-U pattern characteristic of the grey matter. Overall, then, grey matter volume decreases and white matter volume increases in the whole brain, as adolescents mature into adulthood, and the increasing volume of white matter is attributable mainly to late postnatal addition of myelin to the axons [140].

Next, we're interested in narrowing our focus to brain areas important for stress responding. Examining the developmental trajectories of volume changes in these regions might provide important clues for understanding the mechanisms involved in late postnatal plasticity of the adult stress response circuitry. When examining specific brain regions, it has been noted that different regions achieve their peak grey matter volumes at different phases of development, with each region showing its own characteristic developmental trajectory [135]. Unlike grey matter, white matter content increases in a roughly linear fashion, with little overall difference in slope across different areas of the cortex [135]. Thus, we will here consider volume changes in the frontal cortex, which contains the PFC, and also in hippocampus, a subcortical brain structure that plays an important feedback inhibition role during adult HPA axis activation.

As might be predicted based on the pattern of grey matter volume changes observed across development for the whole brain, the grey matter volume in the frontal cortex has been shown to decline with age across ages eight to 30 years [141], indicating that synaptic pruning continues into adulthood in this late-developing region. Cortical white matter, on the other hand, increased across the study period in these same subjects, according to a quadratic function, with the time of greatest change occurring during adolescence [141]. To the best of our knowledge, developmental changes in hippocampal white matter volumes have not been quantified. However, Ostby *et al.* (2009) did quantify hippocampal grey matter volume changes in their sample across eight to 30 years of age, and, surprisingly, this region showed a quadratic increase in grey matter volume up to 30 years of age, with the period of greatest change occurring during adolescence, followed by a leveling off in the twenties age range. This suggests that hippocampal dendrite density continues to increase throughout adolescence, followed by a period of relative stability in early adulthood [141]. This finding likely has important implications for the role of the hippocampus in postnatal development of the stress response. However, a great deal of further work is necessary to elucidate the functional impact of structural change in the hippocampus and other stress-related brain regions during development.

A special MRI technique, called functional MRI (fMRI), was developed not long after the appearance of the original MRI studies in the scientific literature. This technique allows for neuronal activation to be indirectly quantified. For example, brain scans may occur during task performance or in a context with a stressor present, allowing researchers to investigate the involvement of various brain regions in responding to these situations. There are some important limitations associated with the data collection procedures, such as the requirement that subjects remain motionless while being scanned, as for ordinary MRI, which can be a big challenge when working with infants and children and places serious constraints on the kinds of tasks that can be accomplished inside a scanner. Despite these challenges, fMRI has been a revolutionary step in making sense of relationships between brain activity and thoughts, emotions, or behavioural action and will continue to play a prominent role in developmental cognitive neuroscience and related research areas.

4.4.2 Functional Changes

Due to its more recent application and also its recognized limitations, there are less available fMRI data describing changes during brain development, as compared with structural MRI data. Instead of measuring volume, fMRI measures hemodynamic changes associated with neuronal activation in the brain, using a blood oxygen level dependent (BOLD) signal. When neurons fire, they utilize glucose to produce the energy necessary to bring them back to their resting state. Glucose is not stored in the brain, but instead must be transported there in the blood. Thus, neuronal activation leads to an increase in oxygenated blood, relative to deoxygenated blood, in the activated area ~1–2 s following the activation. The contrast between oxygenated and deoxygenated blood can then be visualized to assess the degree of activation in each region of interest.

The approach that is usually taken when using fMRI to study brain development is to passively expose subjects to stimuli or have them perform a mental task while inside the scanner and then compare the consequent neuronal activation in regions of interest among individuals at different ages. Some tasks that require a small amount of motion, such as a brief verbal response, can be accomplished successfully while being scanned; however, in these cases, the study design generally includes measures to limit gross movements and statistically correct for fine movements associated with the vocalization (e.g., [142]). In a cross-sectional design, groups of people in different age brackets are directly compared with each other, whereas, in a longitudinal design, the same individuals are scanned at different ages.

Overall, what has been gleaned from these studies is that, as development progresses, the pattern of response to the experimental manipulation (i.e., stimulus exposure or task performance) changes in various regions of the associated networks, such that some of the involved regions get recruited less with age, while other regions are more heavily recruited with age. It is rarely the case that a brain region previously not involved in the response to a particular manipulation joins the network later in development. Rather, there seems to be a shifting of the relative importance of different brain regions, as the involved circuit becomes more proficient in handling the given experimental situation [131].

One study found that, during passive viewing of human faces, children and adults both recruit brain regions known to be involved in adult face processing, such as the fusiform face area (FFA) [143]. In the children, however, the activation in the FFA wasn't specific to viewing faces but could also be induced by viewing other objects, which wasn't the case in the adults [143]. It appears that the increasing specificity of the FFA response to faces emerges over the adolescent period. A similar result was obtained in a study that tested the neural response to viewing biological motion in groups of children at different ages, with activation in the superior temporal sulcus (STS) becoming increasingly specific to biological motion with age [144]. These studies provide important insight on the development of the 'social brain' (i.e., the neural networks that process socially-relevant stimuli and produce appropriate behavioural responses).

Functional MRI has also been used to study the development of cognition in humans. Cognition is a group of abilities possessed by humans that allows us to think consciously. In involves selective attention, memory, inhibitory control, and language processing, among other abilities. A number of these abilities are designated as 'executive functions,' and research in various fields supports the notion that the PFC is primarily responsible for their emergence. Several fMRI studies demonstrate differential activation in the PFC between children and adults while they conduct tasks assessing executive functions [131]. Some of the studies showing differential recruitment of PFC at different ages involved having the subjects perform 'theory of mind' or 'mentalizing' tasks while in the scanner. This means that the tasks required the subjects to consider the thoughts or intentions of someone else. These studies have consistently shown that the inside part of the PFC (medial PFC; Fig. 7) is much more heavily recruited in children, relative to adults, during performance of these tasks, and, as for the socially-relevant tasks described above, the change appears to occur across the adolescent period [131].

The outside part of the PFC (lateral PFC), on the other hand, has been investigated in regard to its activation during tasks testing other executive functions, such as working memory or inhibitory control. During performance of a task measuring selective attention and inhibitory control (i.e., the Stroop task), activation in the dorsolateral PFC (DLPFC; the upper, outer portion) was greater in adults than in adolescents [145], demonstrating continued functional development of this region even late into the adolescent phase. Interestingly, another research group found that the DLPFC was recruited more heavily in children than in adults in a different test of inhibitory control—the go/no go task [146]. In a different study, the ventrolateral PFC (VLPFC; the lower, outer portion) was activated during performance of the go/no go task in adults, but not in children [147], demonstrating specialization of subregions within the lateral PFC and the potential for different developmental trajectories among these subregions.

Together, these fMRI studies provide information on the development of cognition and associated brain activity in humans. The development of neural circuits involved in stress responding has been much less studied using fMRI, and the emergence of cognitive control over stress responding has not been investigated in this paradigm, to our knowledge. Functional MRI promises to be highly useful for designing future experiments to address various research questions in these areas.

Due to ethical issues, it is very difficult to study stress responding in human participants, especially children. However, carefully designed studies using animal models have provided a wealth of information on the basic neural circuitry involved in stress responding and secretion of the major stress hormones (see Section 1.2.2). The HPA axis has thus come to be known as the major mammalian stress response axis. Its activation by a stressful stimulus results in the release of GCs into the bloodstream, and, accordingly, an elevation in GC levels (i.e., cortisol in humans or corticosterone in rodents) is one of the most routinely used biomarkers of stress responding. Stimuli can be classed as stressors or not stressful for humans, depending on whether or not they induce an increase in cortisol secretion. Using this criterion, researchers have been able to characterize human responses to various stressors, in terms of changes to behaviour and changes in levels of various stress-related hormones. Within a number of clinical populations, groups of people have been identified who have been exposed repeatedly to stressors, such as cases of childhood neglect or abuse. These people have been studied retrospectively, in order to determine the long-term consequences of the repeated stressor exposure. Although highly informative, none of these studies involve a direct examination of neural activation in the stress response circuitry during stressor exposure. Functional MRI provides a means to achieve this, so long as the ethical and technical challenges surrounding stressor exposure in human participants can be adequately surmounted in the experimental design of the study. For example, confinement within an MRI scanner in and of itself can be stressful [148]! With careful planning, such challenges can be overcome, but these studies must be interpreted with caution.

Despite the inherent challenges, there are a few elegant stressor paradigms that have been designed for use in human fMRI studies. One paradigm involves administering an incrementally escalating series of electric shocks to each subject's abdomen until he or she reports that the shock intensity is aversive but still tolerable [149]. Each subject can then be scanned while under the threat of an aversive abdominal shock. This procedure induces neural activation in 'anticipatory stress' circuits and can thus be likened to other human psychological stressors of an anticipatory nature. Studies such as these provide evidence for neural activation in various cortical and subcortical structures known through other means to be involved in mammalian stress responding, such as within the PFC and hippocampus. There are still many pieces of the puzzle missing, in terms of our understanding of brain development in general, and especially development of complex cognitive functions. However, it seems likely that as we gain more insight into the processes that govern postnatal brain development, we will begin to better understand the impact of cortical projections to the HPA axis, and in particular, projections from prefrontal cortical regions that are important for inhibitory control. Studies designed to examine neural activation in these brain regions during stressor exposure at different points across development will be very fruitful in this regard.

4.4.3 Neurotransmitter Systems

Armed with background knowledge that has been gleaned from animal models and post-mortem studies, scientists have a fairly clear concept of what constitutes the basic neuroanatomy of the mammalian stress response system; they've delineated what structures shoulder the biggest responsibility in mediating an effective bodily response when a stressor is encountered (i.e., the HPA axis), and they have also figured out what signaling molecules are most important in this pathway, as well as the typical sequence of their release during stressor exposure. MRI studies involving stress have added significantly to our understanding of HPA axis function in humans and of the protracted development of some structures of the adult stress response system. However, as described in Section 3.3.2, the BOLD signal used in fMRI studies is only an indirect measure of neural activation. There is still no means to directly assess release of neurotransmitters (the signaling molecules released by neural activation; see Section 2.1) in the human brain. Hormone molecules circulating in the body's periphery can be assessed through blood sampling, but the neurotransmitter levels remain elusive. Consequently, scientists have had to turn back to animal models to gather a detailed understanding of the regulation of the HPA axis at the level of neurotransmitter signaling molecules.

In addition to the stress-related neurohormone molecules directly involved in HPA axis activation (see Fig. 3), all of the major mammalian neurotransmitter signaling molecules are involved in some way in stress responding! This includes the small, amino acid transmitters glutamate and gamma-aminobutyric acid (GABA), as well as the monoamines norepinephrine, serotonin, and dopamine. In addition to these classical neurotransmitters, there are other neurotransmitter molecules that are less well studied but that appear to play an important role in stress responding, such as the endocannabinoids [150, 151]. Thus, in order to better understand postnatal development of the mammalian stress response system, we will summarize what is known on development of some of the major neurotransmitter systems involved.

Importantly, the effect that a neurotransmitter (analogous to a key) will have on a particular postsynaptic cell is dependent on which receptor subtype (analogous to a lock) is present on that cell. All of the major neurotransmitters can bind to more than one receptor subtype, and some receptors will make it more difficult for the cell to fire (inhibitory effects on neural activation and subsequent neurotransmitter release), while others will make it easier (excitatory effects on neural activation and subsequent neurotransmitter release). Thus, developmental changes in levels of both the neurotransmitters themselves as well as their receptors could make important contributions to developmental changes in stress responding. Furthermore, the environment modulates some of the developmental changes through epigenetic mechanisms (e.g., [152]). This developmental flexibility allows for the emergence of a phenotype best suited for the specific environmental context, from within a range of potential phenotypes.

Not all of the neurotransmitter systems have been examined thoroughly, in terms of developmental changes in primates. Changes in receptor binding across development have been quantified for most of the major systems in rodents. Most of the evidence gathered to date suggests that receptor levels, rather than levels of neurotransmitters themselves, change across development. Of interest, the 'rise and fall' pattern of change that occurs for other aspects of neural architecture, like grey matter volume and synapse density, is also observed for receptor levels. Each receptor type shows its own characteristic pattern, in terms of the slope of the increase, the age at which the receptor levels peak, and the density at which they stabilize in early adulthood. There are also regional differences in the developmental trajectories for each receptor type. Levels of cortical receptors for glutamate, for example, increase dramatically in early life, reach a peak approximately ten times the levels observed in adults, and then drop off quickly thereafter [131, 153]. Cortical GABA receptor levels follow a similar pattern, although the overall changes in levels are not as drastic [131, 154]. Also, it has been shown that GABA levels can be influenced by the extent of sensory experience [131, 155].

Dopamine (DA) appears to play a particularly important role in modulating stress responding. Specifically, dopamine release in PFC has been associated with cortical modulation of HPA activity, and, furthermore, this stress-evoked PFC activation is asymmetrical, with more activation in the right PFC [156], suggesting that, as the PFC matures, the two sides become specialized for different functions (lateralization of function; [157]). Given the roles of adult PFC regions in mediating cognitive abilities, this suggests that there is a developmental refinement of the processes involved in conscious appraisal of a stressful situation, allowing for a better ability to inhibit physiological bodily responses that are not deemed necessary. Since the DA system plays a prominent role in mediating stress responding in the adult, we will focus on developmental changes observed in levels of prefrontal DA receptors. The 'rise and fall' pattern of expression has been observed for levels of DA receptors in both the striatum ([158]; a subcortical brain structure) and the PFC [159] of rats, although in this species, it appears to be more prominent in males, with females showing a less drastic increase, followed by a leveling off [160]. Furthermore, this sex difference arises independent of pubertal hormone levels [160]. In comparison with striatum, the rise and subsequent fall in DA receptor levels in PFC was more gradual and protracted [159], which is in line with the idea that the PFC is one of the latest developing brain regions. Also, levels of catechol-O-methyltransferase (COMT) activity increase in the PFC from early life to adulthood [161]. This indicates an increase in DA use in this area across development, since this is the primary enzyme responsible for metabolizing DA after it is released from activated neurons. Together, these findings demonstrate that a lot of developmental change occurs in the DA neurotransmitter system, particularly within the PFC, across adolescent development, coinciding with an increased ability in adolescents to inhibit HPA activity through cortical control. It has also been shown that the role of some DA receptor subtypes in stress responding is influenced by experience, such as the role of the D3 or the D4 DA receptor [162]. This implies that there is some degree of developmental flexibility in adolescent development of the DA system. This, combined with the involvement of prefrontal DA signaling in adult stress responding, suggests this is likely to be a fruitful area of research for understanding the effects of adolescent development on the human stress response system.

• • • •

CHAPTER 5

Understanding Adult Stress Responding Using a Developmental Framework

5.1 DISCUSSION

Effective responding to environmental perturbation is key to survival for all organisms—these responses need to be accurate (correct response for the specific perturbation) and timely, but, because such responses are energetically costly, they also need to be discriminating in terms of when they are used, and discrete in terms of how long they last. As we have discussed throughout this book, responses to events that threaten homeostasis (i.e., stressors) are not hard-wired in our genome. Instead, the environment in which the responses will ultimately be needed affects developmental expression of genes, and it is this gene expression that underlies the emergence of the adult phenotype (see Fig. 1). The adult stress response displays large individual differences. Variables such as the nature of the stressor (severity, duration), the effectiveness of terminating the response, and the magnitude of the response are associated with characterizing individuals along a resilient—vulnerable continuum, when it comes to negative outcomes of stress responding. Much work has focused on the development of a vulnerable phenotype, because of the increased propensity of such individuals to develop neuropsychiatric and other diseases. However, there is also increasing interest in discovering the underpinnings of a resilient phenotype, and one concept to emerge is the idea of 'stress inoculation' during early life. Where an individual ends up on this continuum in adulthood is a product of the interaction between genetics (ex. alleles; polymorphisms) and environment (mild vs. severe stress, nutrition, etc.; see Fig. 1).

As discussed in detail and summarized in Figure 1, different components of the major stress response system are said to come 'on-line' at different points during development. An examination of the fMRI data indicates that this is likely to involve an increasing response specificity of particular brain regions, paired with elimination (pruning) of short-range connections within a region or among adjacent regions and strengthening of long-range connections among disparate regions that comprise the functional neural network in the adult. During those times, the brain regions and neurotransmitter systems under development are sensitive to internal and external environmental cues. We focused here on the adolescent period, during which the highest level of control over stress responding ('cognitive' control afforded by the cortex) is programmed. Development of the PFC and key neurotransmitter systems, like the DA system, during this sensitive period occurs on top of earlier gene expression changes within subcortical brain regions, such as the hypothalamus and hippocampus (see Fig. 1), thereby building 'layers' of stress response control from more primitive to more cognitive across development. What is unclear at this point is how the different layers interact, and more work is encouraged that follows individuals from birth (or before) through to adulthood. We touched on some research in section 3.2 that suggests resilience (established by mild early life stress) is associated with better PFC functioning, but more of this type of work, extending to examine the consequences of differing amounts/severity of stress in adulthood is needed. Animal models of the same are also necessary, so that causal relationships and underlying mechanisms can be established. We need to continue to make solid progress in understanding the epigenetics underlying stress response programming, using animal models, while moving forward with innovative research from exciting new fields in human research, such as neuroimaging. For example, there is a large movement devoted to characterizing the connections between regions of the human brain—referred to as the 'map of the human connectome.' This information will provide a point-to-point map of neural pathway connections in the human brain [163]. One can imagine how useful it will be in the future to see how different connections are between 'resilient' and 'vulnerable' individuals in key stress-responsive brain regions. Add to that information regarding epigenetic mechanisms gleaned from animal studies and applied to humans, and we will eventually have a more complete picture of how the stress response develops and is regulated in adulthood. This might allow us to conceive of way that we might help individuals move closer to the resilient end of the continuum, in order to improve their overall health and well being.

5.2 CONCLUSION

As the final stage during which components of the adult stress response phenotype are programmed, the adolescent period plays an important contribution. Unlike the prenatal period and early postnatal periods, when parents are vitally important parts of the external environment, and their actions shape aspects of offspring stress responding, the adolescent period is dominated by a movement away from the parental influence. Adolescence is a time of attaining independence but also of increased sociality with peers, and both of these aspects of the adolescent environment are involved in shaping development of the PFC, and, ultimately, the adult stress response. By this time, development of a resilient versus vulnerable phenotype will be well underway with respect to neural connections and epigenetics that underlie HPA axis output and will be heavily influenced by early parental care; however, cognitive appraisal of a stressful situation, as afforded by connections to and from the PFC and epigenetics of systems involved in higher cognition, will still be malleable, making the adolescent period critically important for determining the adult stress response phenotype and where someone ends up along the resilient—vulnerable continuum.

• • • •

Bibliography

[1] C.H. Waddington, *The Evolution of an Evolutionist*, Cornell University Press, New York, 1975.

[2] S. Yamanaka, Elite and stochastic models for induced pluripotent stem cell generation. *Nature* 460 (2009) 49–52.

[3] C.P. Johnstone, R.D. Reina, and A. Lill, Interpreting indices of physiological stress in free-living vertebrates. *J Comp Physiol B* 182 (2012) 861–79.

[4] M. Joels, G. Fernandez, and B. Roozendaal, Stress and emotional memory: a matter of timing. *Trends Cogn Sci* 15 (2011) 280–8.

[5] P.A. Spadaro, and T.W. Bredy, Emerging role of non-coding RNA in neural plasticity, cognitive function, and neuropsychiatric disorders. *Front Genet* 3 (2012) 132.

[6] C. Tsigos, and G.P. Chrousos, Hypothalamic-pituitary-adrenal axis, neuroendocrine factors and stress. *J Psychosom Res* 53 (2002) 865–71.

[7] E.R. de Kloet, M. Joels, and F. Holsboer, Stress and the brain: from adaptation to disease. *Nat Rev Neurosci* 6 (2005) 463–75.

[8] E.R. De Kloet, W. Sutanto, N. Rots, A. van Haarst, D. van den Berg, M. Oitzl, A. van Eekelen, and D. Voorhuis, Plasticity and function of brain corticosteroid receptors during aging. *Acta Endocrinol (Copenh)* 125 Suppl 1 (1991) 65–72.

[9] R. Jankord, and J.P. Herman, Limbic regulation of hypothalamo-pituitary-adrenocortical function during acute and chronic stress. *Ann N Y Acad Sci* 1148 (2008) 64–73.

[10] J.P. Herman, W.E. Cullinan, M.I. Morano, H. Akil, and S.J. Watson, Contribution of the ventral subiculum to inhibitory regulation of the hypothalamo-pituitary-adrenocortical axis. *J Neuroendocrinol* 7 (1995) 475–82.

[11] N.K. Mueller, C.M. Dolgas, and J.P. Herman, Stressor-selective role of the ventral subiculum in regulation of neuroendocrine stress responses. *Endocrinology* 145 (2004) 3763–8.

[12] S. Levine, Developmental determinants of sensitivity and resistance to stress. *Psychoneuroendocrinology* 30 (2005) 939–46.

[13] D.C. Blanchard, J.K. Shepherd, A. De Padua Carobrez, and R.J. Blanchard, Sex effects in defensive behavior: baseline differences and drug interactions. *Neurosci Biobehav Rev* 15 (1991) 461–8.

[14] D.C. Blanchard, R.L. Spencer, S.M. Weiss, R.J. Blanchard, B. McEwen, and R.R. Sakai, Visible burrow system as a model of chronic social stress: behavioral and neuroendocrine correlates. *Psychoneuroendocrinology* 20 (1995) 117–34.

[15] H. Ylonen, Vole cycles and antipredatory behaviour. *Trends Ecol Evol* 9 (1994) 426–30.

[16] M. Moghimian, M. Faghihi, S.M. Karimian, and A. Imani, The effect of acute stress exposure on ischemia and reperfusion injury in rat heart: role of oxytocin. *Stress* 15 (2012) 385–92.

[17] R.D. Porsolt, G. Anton, N. Blavet, and M. Jalfre, Behavioural despair in rats: a new model sensitive to antidepressant treatments. *Eur J Pharmacol* 47 (1978) 379–91.

[18] L.K. Takahashi, B.R. Nakashima, H. Hong, and K. Watanabe, The smell of danger: a behavioral and neural analysis of predator odor-induced fear. *Neurosci Biobehav Rev* 29 (2005) 1157–67.

[19] L.D. Wright, K.E., and T.S. Perrot, Stress responses of adolescent male and female rats exposed repeatedly to cat odor stimuli, and long-term enhancement of adult defensive behaviors. *Dev Psychobiol* (2012).

[20] L.D. Wright, K.E. Muir, and T.S. Perrot, Enhanced stress responses in adolescent versus adult rats exposed to cues of predation threat, and peer interaction as a predictor of adult defensiveness. *Dev Psychobiol* 54 (2012) 47–69.

[21] L.D. Wright, K.E. Hebert, and T.S. Perrot-Sinal, Periadolescent stress exposure exerts long-term effects on adult stress responding and expression of prefrontal dopamine receptors in male and female rats. *Psychoneuroendocrinology* 33 (2008) 130–42.

[22] R. Mashoodh, L.D. Wright, K. Hebert, and T.S. Perrot-Sinal, Investigation of sex differences in behavioural, endocrine, and neural measures following repeated psychological stressor exposure. *Behav Brain Res* 188 (2008) 368–79.

[23] H.F. Figueiredo, C.M. Dolgas, and J.P. Herman, Stress activation of cortex and hippocampus is modulated by sex and stage of estrus. *Endocrinology* 143 (2002) 2534–40.

[24] K.A. Matthews, C.R. Katholi, H. McCreath, M.A. Whooley, D.R. Williams, S. Zhu, and J.H. Markovitz, Blood pressure reactivity to psychological stress predicts hypertension in the CARDIA study. *Circulation* 110 (2004) 74–8.

[25] K.A. Matthews, S. Zhu, D.C. Tucker, and M.A. Whooley, Blood pressure reactivity to psychological stress and coronary calcification in the Coronary Artery Risk Development in Young Adults Study. *Hypertension* 47 (2006) 391–5.

[26] K.S. Kendler, R.C. Kessler, E.E. Walters, C. MacLean, M.C. Neale, A.C. Heath, and L.J. Eaves, Stressful life events, genetic liability, and onset of an episode of major depression in women. *Am J Psychiatry* 152 (1995) 833–42.

[27] B.M. Kudielka, and C. Kirschbaum, Sex differences in HPA axis responses to stress: a review. *Biol Psychol* 69 (2005) 113–32.

[28] E. Kajantie, and D.I. Phillips, The effects of sex and hormonal status on the physiological response to acute psychosocial stress. *Psychoneuroendocrinology* 31 (2006) 151–78.

[29] C. Dalla, P.M. Pitychoutis, N. Kokras, and Z. Papadopoulou-Daifoti, Sex differences in animal models of depression and antidepressant response. *Basic Clin Pharmacol Toxicol* 106 (2010) 226–33.

[30] C. Holden, Sex and the suffering brain. *Science* 308 (2005) 1574.

[31] B. Lenz, C.P. Muller, C. Stoessel, W. Sperling, T. Biermann, T. Hillemacher, S. Bleich, and J. Kornhuber, Sex hormone activity in alcohol addiction: integrating organizational and activational effects. *Prog Neurobiol* 96 (2012) 136–63.

[32] R.C. Kessler, Epidemiology of women and depression. *J Affect Disord* 74 (2003) 5–13.

[33] R.C. Kessler, K.A. McGonagle, S. Zhao, C.B. Nelson, M. Hughes, S. Eshleman, H.U. Wittchen, and K.S. Kendler, Lifetime and 12-month prevalence of DSM-III-R psychiatric disorders in the United States. Results from the National Comorbidity Survey. *Arch Gen Psychiatry* 51 (1994) 8–19.

[34] Z. Petanjek, M. Judas, G. Simic, M.R. Rasin, H.B. Uylings, P. Rakic, and I. Kostovic, Extraordinary neoteny of synaptic spines in the human prefrontal cortex. *Proc Natl Acad Sci U S A* 108 (2011) 13281–6.

[35] C. Heim, L.J. Young, D.J. Newport, T. Mletzko, A.H. Miller, and C.B. Nemeroff, Lower CSF oxytocin concentrations in women with a history of childhood abuse. *Mol Psychiatry* 14 (2009) 954–8.

[36] R.F. Anda, V.J. Felitti, J.D. Bremner, J.D. Walker, C. Whitfield, B.D. Perry, S.R. Dube, and W.H. Giles, The enduring effects of abuse and related adverse experiences in childhood. A convergence of evidence from neurobiology and epidemiology. *Eur Arch Psychiatry Clin Neurosci* 256 (2006) 174–86.

[37] R.F. Anda, D.W. Brown, V.J. Felitti, S.R. Dube, and W.H. Giles, Adverse childhood experiences and prescription drug use in a cohort study of adult HMO patients. *BMC Public Health* 8 (2008) 198.

[38] R.F. Anda, D.W. Brown, S.R. Dube, J.D. Bremner, V.J. Felitti, and W.H. Giles, Adverse childhood experiences and chronic obstructive pulmonary disease in adults. *Am J Prev Med* 34 (2008) 396–403.

[39] C. Heim, D.J. Newport, T. Mletzko, A.H. Miller, and C.B. Nemeroff, The link between childhood trauma and depression: insights from HPA axis studies in humans. *Psychoneuroendocrinology* 33 (2008) 693–710.

[40] R.L. Repetti, S.E. Taylor, and T.E. Seeman, Risky families: family social environments and the mental and physical health of offspring. *Psychol Bull* 128 (2002) 330–66.

[41] E. McCrory, S.A. De Brito, and E. Viding, The impact of childhood maltreatment: a review of neurobiological and genetic factors. *Front Psychiatry* 2 (2011) 48.

[42] R.J. Turner, The pursuit of socially modifiable contingencies in mental health. *J Health Soc Behav* 44 (2003) 1–17.

[43] M. Rutter, Implications of resilience concepts for scientific understanding. *Ann NY Acad Sci* 1094 (2006) 1–12.

[44] D.M. Khoshaba, and S.R. Maddi, Early experiences in hardiness development. *Consult Psychol J* 51 (1999) 106–16.

[45] M. Katz, C. Liu, M. Schaer, K.J. Parker, M.C. Ottet, A. Epps, C.L. Buckmaster, R. Bammer, M.E. Moseley, A.F. Schatzberg, S. Eliez, and D.M. Lyons, Prefrontal plasticity and stress inoculation-induced resilience. *Dev Neurosci* 31 (2009) 293–9.

[46] M.L. Lehmann, and M. Herkenham, Environmental enrichment confers stress resiliency to social defeat through an infralimbic cortex-dependent neuroanatomical pathway. *J Neurosci* 31 (2011) 6159–73.

[47] R.R. Rozeske, A.K. Evans, M.G. Frank, L.R. Watkins, C.A. Lowry, and S.F. Maier, Uncontrollable, but not controllable, stress desensitizes 5-HT1A receptors in the dorsal raphe nucleus. *J Neurosci* 31 (2011) 14107–15.

[48] S. Levine, Infantile experience and resistance to physiological stress. *Science* 126 (1957) 405.

[49] G.N. Levine, Anxiety about illness: psychological and social bases. *J Health Hum Behav* 3 (1962) 30–4.

[50] M.J. Meaney, S. Bhatnagar, S. Larocque, C. McCormick, N. Shanks, S. Sharma, J. Smythe, V. Viau, and P.M. Plotsky, Individual differences in the hypothalamic-pituitary-adrenal stress response and the hypothalamic CRF system. *Ann NY Acad Sci* 697 (1993) 70–85.

[51] P.M. Plotsky, and M.J. Meaney, Early, postnatal experience alters hypothalamic corticotropin-releasing factor (CRF) mRNA, median eminence CRF content and stress-induced release in adult rats. *Brain Res Mol Brain Res* 18 (1993) 195–200.

[52] D.M. Lyons, K.J. Parker, and A.F. Schatzberg, Animal models of early life stress: Implications for understanding resilience. *Dev Psychobiol* 52 (2010) 402–10.

[53] S. Levine, and T. Mody, The long-term psychobiological consequences of intermittent postnatal separation in the squirrel monkey. *Neurosci Biobehav Rev* 27 (2003) 83–9.

[54] D.M. Lyons, O.J. Wang, S.E. Lindley, S. Levine, N.H. Kalin, and A.F. Schatzberg, Separation induced changes in squirrel monkey hypothalamic-pituitary-adrenal physiology resemble aspects of hypercortisolism in humans. *Psychoneuroendocrinology* 24 (1999) 131–42.

[55] D.M. Lyons, C. Yang, B.W. Mobley, J.T. Nickerson, and A.F. Schatzberg, Early environmental regulation of glucocorticoid feedback sensitivity in young adult monkeys. *J Neuroendocrinol* 12 (2000) 723–8.

[56] D.F. Hawley, M. Bardi, A.M. Everette, T.J. Higgins, K.M. Tu, C.H. Kinsley, and K.G. Lambert, Neurobiological constituents of active, passive, and variable coping strategies in

rats: integration of regional brain neuropeptide Y levels and cardiovascular responses. *Stress* 13 (2010) 172–83.

[57] S.J. Spencer, K.M. Buller, and T.A. Day, Medial prefrontal cortex control of the paraventricular hypothalamic nucleus response to psychological stress: possible role of the bed nucleus of the stria terminalis. *J Comp Neurol* 481 (2005) 363–76.

[58] J.J. Radley, K.L. Gosselink, and P.E. Sawchenko, A discrete GABAergic relay mediates medial prefrontal cortical inhibition of the neuroendocrine stress response. *J Neurosci* 29 (2009) 7330–40.

[59] S.J. McDougall, R.E. Widdop, and A.J. Lawrence, Medial prefrontal cortical integration of psychological stress in rats. *Eur J Neurosci* 20 (2004) 2430–40.

[60] M.D. Lapiz-Bluhm, A.E. Soto-Pina, J.G. Hensler, and D.A. Morilak, Chronic intermittent cold stress and serotonin depletion induce deficits of reversal learning in an attentional set-shifting test in rats. *Psychopharmacology (Berl)* 202 (2009) 329–41.

[61] G.A. Bonanno, A. Papa, K. Lalande, M. Westphal, and K. Coifman, The importance of being flexible: the ability to both enhance and suppress emotional expression predicts long-term adjustment. *Psychol Sci* 15 (2004) 482–7.

[62] S.B. Floresco, and O. Magyar, Mesocortical dopamine modulation of executive functions: beyond working memory. *Psychopharmacology (Berl)* 188 (2006) 567–85.

[63] J.J. Cerqueira, O.F. Almeida, and N. Sousa, The stressed prefrontal cortex. Left? Right! *Brain Behav Immun* (2008).

[64] K. Matsuo, M. Nicoletti, K. Nemoto, J.P. Hatch, M.A. Peluso, F.G. Nery, and J.C. Soares, A voxel-based morphometry study of frontal gray matter correlates of impulsivity. *Hum Brain Mapp* 30 (2009) 1188–95.

[65] K.J. Parker, C.L. Buckmaster, K.R. Justus, A.F. Schatzberg, and D.M. Lyons, Mild early life stress enhances prefrontal-dependent response inhibition in monkeys. *Biol Psychiatry* 57 (2005) 848–55.

[66] F.A. Champagne, Epigenetic mechanisms and the transgenerational effects of maternal care. *Front Neuroendocrinol* 29 (2008) 386–97.

[67] D. Liu, J. Diorio, B. Tannenbaum, C. Caldji, D. Francis, A. Freedman, S. Sharma, D. Pearson, P.M. Plotsky, and M.J. Meaney, Maternal care, hippocampal glucocorticoid receptors, and hypothalamic-pituitary-adrenal responses to stress. *Science* 277 (1997) 1659–62.

[68] H. Wurbel, Ideal homes? Housing effects on rodent brain and behaviour. *Trends Neurosci* 24 (2001) 207–11.

[69] C. Caldji, B. Tannenbaum, S. Sharma, D. Francis, P.M. Plotsky, and M.J. Meaney, Maternal care during infancy regulates the development of neural systems mediating the expression of fearfulness in the rat. *Proc Natl Acad Sci U S A* 95 (1998) 5335–40.

[70] F.A. Champagne, D.D. Francis, A. Mar, and M.J. Meaney, Variations in maternal care in the rat as a mediating influence for the effects of environment on development. *Physiol Behav* 79 (2003) 359–71.

[71] D.D. Francis, F.A. Champagne, D. Liu, and M.J. Meaney, Maternal care, gene expression, and the development of individual differences in stress reactivity. *Ann N Y Acad Sci* 896 (1999) 66–84.

[72] D.A. Gutman, and C.B. Nemeroff, Neurobiology of early life stress: rodent studies. *Semin Clin Neuropsychiatry* 7 (2002) 89–95.

[73] D.D. Francis, and M.J. Meaney, Maternal care and the development of stress responses. *Curr Opin Neurobiol* 9 (1999) 128–34.

[74] M.J. Meaney, Maternal care, gene expression, and the transmission of individual differences in stress reactivity across generations. *Annu Rev Neurosci* 24 (2001) 1161–92.

[75] C.R. Pryce, and J. Feldon, Long-term neurobehavioural impact of the postnatal environment in rats: manipulations, effects and mediating mechanisms. *Neurosci Biobehav Rev* 27 (2003) 57–71.

[76] S. Macri, G.J. Mason, and H. Wurbel, Dissociation in the effects of neonatal maternal separations on maternal care and the offspring's HPA and fear responses in rats. *Eur J Neurosci* 20 (2004) 1017–24.

[77] F. Cirulli, A. Berry, and E. Alleva, Early disruption of the mother-infant relationship: effects on brain plasticity and implications for psychopathology. *Neurosci Biobehav Rev* 27 (2003) 73–82.

[78] I.C. Weaver, Epigenetic programming by maternal behavior and pharmacological intervention. Nature versus nurture: let's call the whole thing off. *Epigenetics* 2 (2007) 22–8.

[79] J. McLeod, C.J. Sinal, and T.S. Perrot-Sinal, Evidence for non-genomic transmission of ecological information via maternal behavior in female rats. *Genes Brain Behav* 6 (2007) 19–29.

[80] R. Mashoodh, C.J. Sinal, and T.S. Perrot-Sinal, Predation threat exerts specific effects on rat maternal behaviour and anxiety-related behaviour of male and female offspring. *Physiol Behav* 96 (2009) 693–702.

[81] A.C. Tang, K.G. Akers, B.C. Reeb, R.D. Romeo, and B.S. McEwen, Programming social, cognitive, and neuroendocrine development by early exposure to novelty. *Proc Natl Acad Sci U S A* 103 (2006) 15716–21.

[82] H.F. Figueiredo, A. Bruestle, B. Bodie, C.M. Dolgas, and J.P. Herman, The medial prefrontal cortex differentially regulates stress-induced c-fos expression in the forebrain depending on type of stressor. *Eur J Neurosci* 18 (2003) 2357–64.

[83] D. Diorio, V. Viau, and M.J. Meaney, The role of the medial prefrontal cortex (cingulate

gyrus) in the regulation of hypothalamic-pituitary-adrenal responses to stress. *J Neurosci* 13 (1993) 3839–47.

[84] M.M. Sanchez, L.J. Young, P.M. Plotsky, and T.R. Insel, Distribution of corticosteroid receptors in the rhesus brain: relative absence of glucocorticoid receptors in the hippocampal formation. *J Neurosci* 20 (2000) 4657–68.

[85] M.S. Weinberg, D.C. Johnson, A.P. Bhatt, and R.L. Spencer, Medial prefrontal cortex activity can disrupt the expression of stress response habituation. *Neuroscience* 168 (2010) 744–56.

[86] J.J. Radley, C.M. Arias, and P.E. Sawchenko, Regional differentiation of the medial prefrontal cortex in regulating adaptive responses to acute emotional stress. *J Neurosci* 26 (2006) 12967–76.

[87] H.F. Figueiredo, B.L. Bodie, M. Tauchi, C.M. Dolgas, and J.P. Herman, Stress integration after acute and chronic predator stress: differential activation of central stress circuitry and sensitization of the hypothalamo-pituitary-adrenocortical axis. *Endocrinology* 144 (2003) 5249–58.

[88] S. Kern, T.R. Oakes, C.K. Stone, E.M. McAuliff, C. Kirschbaum, and R.J. Davidson, Glucose metabolic changes in the prefrontal cortex are associated with HPA axis response to a psychosocial stressor. *Psychoneuroendocrinology* 33 (2008) 517–29.

[89] K. Ondicova, R. Kvetnansky, and B. Mravec, Medial prefrontal cortex transection enhanced stress-induced activation of sympathoadrenal system in rats. *Endocr Regul* 46 (2012) 129–36.

[90] A.L. Jahn, A.S. Fox, H.C. Abercrombie, S.E. Shelton, T.R. Oakes, R.J. Davidson, and N.H. Kalin, Subgenual prefrontal cortex activity predicts individual differences in hypothalamic-pituitary-adrenal activity across different contexts. *Biol Psychiatry* 67 (2010) 175–81.

[91] S.G. Matthews, and D.I. Phillips, Transgenerational inheritance of stress pathology. *Exp Neurol* 233 (2012) 95–101.

[92] K.M. Gudsnuk, and F.A. Champagne, Epigenetic effects of early developmental experiences. *Clin Perinatol* 38 (2011) 703–17.

[93] R.G. Hunter, Epigenetic effects of stress and corticosteroids in the brain. *Front Cell Neurosci* 6 (2012) 18.

[94] I.C. Weaver, A.C. D'Alessio, S.E. Brown, I.C. Hellstrom, S. Dymov, S. Sharma, M. Szyf, and M.J. Meaney, The transcription factor nerve growth factor-inducible protein a mediates epigenetic programming: altering epigenetic marks by immediate-early genes. *J Neurosci* 27 (2007) 1756–68.

[95] E. McCrory, S.A. De Brito, and E. Viding, Research review: the neurobiology and genetics of maltreatment and adversity. *J Child Psychol Psychiatry* 51 (2010) 1079–95.

[96] N. Tsankova, W. Renthal, A. Kumar, and E.J. Nestler, Epigenetic regulation in psychiatric disorders. *Nat Rev Neurosci* 8 (2007) 355–67.

[97] T.L. Roth, F.D. Lubin, A.J. Funk, and J.D. Sweatt, Lasting epigenetic influence of early-life adversity on the BDNF gene. *Biol Psychiatry* 65 (2009) 760–9.

[98] L.T. Huang, The link between perinatal glucocorticoids exposure and psychiatric disorders. *Pediatr Res* 69 (2011) 19R–25R.

[99] A. Korosi, M. Shanabrough, S. McClelland, Z.W. Liu, E. Borok, X.B. Gao, T.L. Horvath, and T.Z. Baram, Early-life experience reduces excitation to stress-responsive hypothalamic neurons and reprograms the expression of corticotropin-releasing hormone. *J Neurosci* 30 (2010) 703–13.

[100] H. Danker-Hopfe, and K. Delibalta, Menarcheal age of Turkish girls in Bremen. *Anthropol Anz* 48 (1990) 1–14.

[101] S.E. Anderson, G.E. Dallal, and A. Must, Relative weight and race influence average age at menarche: results from two nationally representative surveys of US girls studied 25 years apart. *Pediatrics* 111 (2003) 844–50.

[102] B. Al-Sahab, C.I. Ardern, M.J. Hamadeh, and H. Tamim, Age at menarche in Canada: results from the National Longitudinal Survey of Children & Youth. *BMC Public Health* 10 (2010) 736.

[103] M. Jorgensen, N. Keiding, and N.E. Skakkebaek, Estimation of spermarche from longitudinal spermaturia data. *Biometrics* 47 (1991) 177–93.

[104] V. Abbassi, Growth and normal puberty. *Pediatrics* 102 (1998) 507–11.

[105] R.W. Taylor, A.M. Grant, S.M. Williams, and A. Goulding, Sex differences in regional body fat distribution from pre- to postpuberty. *Obesity (Silver Spring)* 18 (2010) 1410–6.

[106] V. Rochira, L. Zirilli, A.D. Genazzani, A. Balestrieri, C. Aranda, B. Fabre, P. Antunez, C. Diazzi, C. Carani, and L. Maffei, Hypothalamic-pituitary-gonadal axis in two men with aromatase deficiency: evidence that circulating estrogens are required at the hypothalamic level for the integrity of gonadotropin negative feedback. *Eur J Endocrinol* 155 (2006) 513–22.

[107] L.L. Gibson, L. Hahner, S. Osborne-Lawrence, Z. German, K.K. Wu, K.L. Chambliss, and P.W. Shaul, Molecular basis of estrogen-induced cyclooxygenase type 1 upregulation in endothelial cells. *Circ Res* 96 (2005) 518–25.

[108] C.G. Castles, S. Oesterreich, R. Hansen, and S.A. Fuqua, Auto-regulation of the estrogen receptor promoter. *J Steroid Biochem Mol Biol* 62 (1997) 155–63.

[109] A. van de Stolpe, A.J. Slycke, M.O. Reinders, A.W. Zomer, S. Goodenough, C. Behl, A.F. Seasholtz, and P.T. van der Saag, Estrogen receptor (ER)-mediated transcriptional regulation of the human corticotropin-releasing hormone-binding protein promoter: differential effects of ERalpha and ERbeta. *Mol Endocrinol* 18 (2004) 2908–23.

[110] J. Qiu, O.K. Ronnekleiv, and M.J. Kelly, Modulation of hypothalamic neuronal activity through a novel G-protein-coupled estrogen membrane receptor. *Steroids* 73 (2008) 985–91.

[111] C.C. Zouboulis, Acne and sebaceous gland function. *Clin Dermatol* 22 (2004) 360–6.

[112] E. Johnson, Steroids and specialized skin secretions in mammals. *Biochem Soc Trans* 4 (1976) 602–5.

[113] A.D. Rogol, Growth at puberty: interaction of androgens and growth hormone. *Med Sci Sports Exerc* 26 (1994) 767–70.

[114] H. Doneray, and Z. Orbak, Association between anthropometric hormonal measurements and bone mineral density in puberty and constitutional delay of growth and puberty. *West Indian Med J* 59 (2010) 125–30.

[115] L. Xu, Q. Wang, A. Lyytikainen, T. Mikkola, E. Volgyi, S. Cheng, P. Wiklund, E. Munukka, P. Nicholson, and M. Alen, Concerted actions of insulin-like growth factor 1, testosterone, and estradiol on peripubertal bone growth: a 7-year longitudinal study. *J Bone Miner Res* 26 (2011) 2204–11.

[116] A.S. Chagin, and L. Savendahl, Genes of importance in the hormonal regulation of growth plate cartilage. *Horm Res* 71 Suppl 2 (2009) 41–7.

[117] P.J. O'Shaughnessy, A. Monteiro, G. Verhoeven, K. De Gendt, and M.H. Abel, Effect of FSH on testicular morphology and spermatogenesis in gonadotrophin-deficient hypogonadal mice lacking androgen receptors. *Reproduction* 139 (2010) 177–84.

[118] P.J. O'Shaughnessy, A. Monteiro, and M. Abel, Testicular development in mice lacking receptors for follicle stimulating hormone and androgen. *PLoS One* 7 (2012) e35136.

[119] Z. Janczewski, and L. Bablok, Semen characteristics in pubertal boys. I. Semen quality after first ejaculation. *Arch Androl* 15 (1985) 199–205.

[120] Z. Janczewski, and L. Bablok, Semen characteristics in pubertal boys. II. Semen quality in relation to bone age. *Arch Androl* 15 (1985) 207–11.

[121] Z. Janczewski, and L. Bablok, Semen characteristics in pubertal boys. III. Semen quality and somatosexual development. *Arch Androl* 15 (1985) 213–8.

[122] Z. Janczewski, and L. Bablok, Semen characteristics in pubertal boys. IV. Semen quality and hormone profile. *Arch Androl* 15 (1985) 219–23.

[123] M. Kennedy, Hormonal regulation of hepatic drug-metabolizing enzyme activity during adolescence. *Clin Pharmacol Ther* 84 (2008) 662–73.

[124] C.W. Colvin, and H. Abdullatif, Anatomy of female puberty: The clinical relevance of developmental changes in the reproductive system. *Clin Anat* (2012).

[125] W.A. Marshall, and J.M. Tanner, Puberty. in: F. Falkner, and J.M. Tanner, (Eds.), *Human Growth: A Comprehensive Treatise*, Springer, New York, 1986, pp. 171–210.

[126] C.M. Gordon, and M.R. Laufer, The Physiology of Puberty, in S.J. Emans, M.R. Laufer, and D.P. Goldstein, (Eds.), *Pediatric and Adolescent Gynecology*, Lippincott Williams & Wilkins, Philadelphia, 2005, pp. 120–155.

[127] D. Apter, Serum steroids and pituitary hormones in female puberty: a partly longitudinal study. *Clin Endocrinol (Oxf)* 12 (1980) 107–20.

[128] J.M. Tanner, and P.S. Davies, Clinical longitudinal standards for height and height velocity for North American children. *J Pediatr* 107 (1985) 317–29.

[129] P.E. Clayton, and J.A. Trueman, Leptin and puberty. *Arch Dis Child* 83 (2000) 1–4.

[130] R. Boyar, J. Finkelstein, H. Roffwarg, S. Kapen, E. Weitzman, and L. Hellman, Synchronization of augmented luteinizing hormone secretion with sleep during puberty. *N Engl J Med* 287 (1972) 582–6.

[131] M.H. Johnson, *Developmental Cognitive Neuroscience*, Wiley-Blackwell, Chichester, 2011.

[132] V.G. Carrion, and S.S. Wong, Can traumatic stress alter the brain? Understanding the implications of early trauma on brain development and learning. *J Adolesc Health* 51 (2012) S23–8.

[133] R.D. Romeo, Pubertal maturation and programming of hypothalamic-pituitary-adrenal reactivity. *Front Neuroendocrinol* 31 (2010) 232–40.

[134] R.D. Romeo, Adolescence: a central event in shaping stress reactivity. *Dev Psychobiol* 52 (2010) 244–53.

[135] J.N. Giedd, Structural magnetic resonance imaging of the adolescent brain. *Ann N Y Acad Sci* 1021 (2004) 77–85.

[136] M. Venere, Y.G. Han, J.S. Song, R. Bell, A. Alvarez-Buylla, and R. Blelloch, Sox1 marks an activated neural stem/progenitor cell in the hippocampus. *Development* (2012).

[137] M. Zusso, and P. Debetto, Isolation and culture of neural progenitor cells from rat postnatal cerebellum. *Methods Mol Biol* 846 (2012) 39–47.

[138] D.O. Hebb, *The Organization of Behavior*, John Wiley & Sons Inc., New York, 1949.

[139] J.N. Giedd, and J.L. Rapoport, Structural MRI of pediatric brain development: what have we learned and where are we going? *Neuron* 67 (2010) 728–34.

[140] K.M. Welker, and A. Patton, Assessment of normal myelination with magnetic resonance imaging. *Semin Neurol* 32 (2012) 15–28.

[141] Y. Ostby, C.K. Tamnes, A.M. Fjell, L.T. Westlye, P. Due-Tonnessen, and K.B. Walhovd, Heterogeneity in subcortical brain development: A structural magnetic resonance imaging study of brain maturation from 8 to 30 years. *J Neurosci* 29 (2009) 11772–82.

[142] S.A. Gruber, J. Rogowska, P. Holcomb, S. Soraci, and D. Yurgelun-Todd, Stroop performance in normal control subjects: an fMRI study. *Neuroimage* 16 (2002) 349–60.

[143] K.S. Scherf, M. Behrmann, K. Humphreys, and B. Luna, Visual category-selectivity for

faces, places and objects emerges along different developmental trajectories. *Dev Sci* 10 (2007) F15–30.

[144] M.W. Mosconi, P.B. Mack, G. McCarthy, and K.A. Pelphrey, Taking an "intentional stance" on eye-gaze shifts: a functional neuroimaging study of social perception in children. *Neuroimage* 27 (2005) 247–52.

[145] N.E. Adleman, V. Menon, C.M. Blasey, C.D. White, I.S. Warsofsky, G.H. Glover, and A.L. Reiss, A developmental fMRI study of the Stroop color-word task. *Neuroimage* 16 (2002) 61–75.

[146] J.R. Booth, D.D. Burman, J.R. Meyer, Z. Lei, B.L. Trommer, N.D. Davenport, W. Li, T.B. Parrish, D.R. Gitelman, and M.M. Mesulam, Neural development of selective attention and response inhibition. *Neuroimage* 20 (2003) 737–51.

[147] S.A. Bunge, N.M. Dudukovic, M.E. Thomason, C.J. Vaidya, and J.D. Gabrieli, Immature frontal lobe contributions to cognitive control in children: evidence from fMRI. *Neuron* 33 (2002) 301–11.

[148] J. Enders, E. Zimmermann, M. Rief, P. Martus, R. Klingebiel, P. Asbach, C. Klessen, G. Diederichs, T. Bengner, U. Teichgraber, B. Hamm, and M. Dewey, Reduction of claustrophobia during magnetic resonance imaging: methods and design of the "CLAUSTRO" randomized controlled trial. *BMC Med Imaging* 11 (2011) 4.

[149] C.S. Hubbard, J.S. Labus, J. Bueller, J. Stains, B. Suyenobu, G.E. Dukes, D.L. Kelleher, K. Tillisch, B.D. Naliboff, and E.A. Mayer, Corticotropin-releasing factor receptor 1 antagonist alters regional activation and effective connectivity in an emotional-arousal circuit during expectation of abdominal pain. *J Neurosci* 31 (2011) 12491–500.

[150] E.J. Carrier, S. Patel, and C.J. Hillard, Endocannabinoids in neuroimmunology and stress. *Curr Drug Targets CNS Neurol Disord* 4 (2005) 657–65.

[151] W.M. Olango, M. Roche, G.K. Ford, B. Harhen, and D.P. Finn, The endocannabinoid system in the rat dorsolateral periaqueductal grey mediates fear-conditioned analgesia and controls fear expression in the presence of nociceptive tone. *Br J Pharmacol* 165 (2012) 2549–60.

[152] I.C. Weaver, N. Cervoni, F.A. Champagne, A.C. D'Alessio, S. Sharma, J.R. Seckl, S. Dymov, M. Szyf, and M.J. Meaney, Epigenetic programming by maternal behavior. *Nat Neurosci* 7 (2004) 847–54.

[153] R. Schliebs, E. Kullmann, and V. Bigl, Development of glutamate binding sites in the visual structures of the rat brain. Effect of visual pattern deprivation. *Biomed Biochim Acta* 45 (1986) 495–506.

[154] B.W. Brooksbank, D.J. Atkinson, and R. Balazs, Biochemical development of the human brain. II. Some parameters of the GABA-ergic system. *Dev Neurosci* 4 (1981) 188–200.

[155] V.M. Fosse, P. Heggelund, and F. Fonnum, Postnatal development of glutamatergic, GABAergic, and cholinergic neurotransmitter phenotypes in the visual cortex, lateral geniculate nucleus, pulvinar, and superior colliculus in cats. *J Neurosci* 9 (1989) 426–35.

[156] J.N. Carlson, L.W. Fitzgerald, R.W. Keller, Jr., and S.D. Glick, Lateralized changes in prefrontal cortical dopamine activity induced by controllable and uncontrollable stress in the rat. *Brain Res* 630 (1993) 178–87.

[157] J.N. Carlson, L.W. Fitzgerald, R.W. Keller, Jr., and S.D. Glick, Side and region dependent changes in dopamine activation with various durations of restraint stress. *Brain Res* 550 (1991) 313–8.

[158] M.H. Teicher, E. Krenzel, A.P. Thompson, and S.L. Andersen, Dopamine receptor pruning during the peripubertal period is not attenuated by NMDA receptor antagonism in rat. *Neurosci Lett* 339 (2003) 169–71.

[159] S.L. Andersen, A.T. Thompson, M. Rutstein, J.C. Hostetter, and M.H. Teicher, Dopamine receptor pruning in prefrontal cortex during the periadolescent period in rats. *Synapse* 37 (2000) 167–9.

[160] S.L. Andersen, A.P. Thompson, E. Krenzel, and M.H. Teicher, Pubertal changes in gonadal hormones do not underlie adolescent dopamine receptor overproduction. *Psychoneuroendocrinology* 27 (2002) 683–91.

[161] E.M. Tunbridge, C.S. Weickert, J.E. Kleinman, M.M. Herman, J. Chen, B.S. Kolachana, P.J. Harrison, and D.R. Weinberger, Catechol-o-methyltransferase enzyme activity and protein expression in human prefrontal cortex across the postnatal lifespan. *Cereb Cortex* 17 (2007) 1206–12.

[162] B. El-Khodor, and P. Boksa, Caesarean section birth produces long term changes in dopamine D1 receptors and in stress-induced regulation of D3 and D4 receptors in the rat brain. *Neuropsychopharmacology* 25 (2001) 423–39.

[163] A.W. Toga, K.A. Clark, P.M. Thompson, D.W. Shattuck, and J.D. Van Horn, Mapping the human connectome. *Neurosurgery* 71 (2012) 1–5.

Author Biographies

Dr. Lisa Wright completed her PhD in psychology and neuroscience at Dalhousie University, Halifax, Nova Scotia, Canada. Her thesis work for both her Master's (2005) and PhD (2011) degrees was conducted under the supervision of Dr. Tara Perrot, with whom she has studied adolescent brain development, associated behavioural changes, and the effects of repeated stressor exposure, using a rat model. Prior to this, Dr. Wright completed an Honours degree (2002) in biology with a minor in psychology in her hometown of Fredericton, at the University of New Brunswick. Currently, she is doing research work and teaching at Dalhousie and at Acadia University, Wolfville, Nova Scotia.

Dr. Tara Perrot began her academic career by completing a BSc Honours degree in psychology at the University of Western Ontario (UWO), London, Ontario, Canada. It was during this time that she became intrigued with neuroscience research under the tutelage of Dr. Mel Goodale. She continued her career at UWO, entering the Neuroscience Graduate Program and beginning her research dedicated to understanding sex differences in stress responding, culminating in a PhD in 1998. Dr. Perrot was intent on expanding her research skills and moved to the University of Maryland School of Medicine to pursue a Natural Sciences and Engineering Research Council of Canada Postdoctoral Fellowship with Dr. Margaret McCarthy. In 2002, Dr. Perrot joined the Faculty of the Department of Psychology and Neuroscience at Dalhousie University, Halifax, Nova Scotia, Canada. She is an associate professor in that department and in the Department of Anatomy and Neurobiology. Dr. Perrot's present research interests are focused on understanding the influence of internal and external cues during development in shaping the adult stress response.

